Brief history of

EARTH AND ENERGY, POWER AND POLLUTION

The historic natural climate is not changing ...

...we are polluting the earth.

PATRICK SIMMONS

AUTHOR CABIN
THE PLACE FOR YOUR STORY

Resting in Peace

Arlene M. Simmons
1936-1989

Suzanne E. Simmons
1956-2018

Douglas J. Simmons
1958-2011

PREFACE

What the Book Is About and Answers To Questions

Universe: All matter, time, and energy in boundless space.

Earth: A planet within a solar system in the universe.

Energy: Ability to perform work - chemical and the atmosphere.

Power: Rate of energy applied to perform work.

Pollution: Unused or inefficient power and chemical waste.

Climate: Weather conditions based on earth's position relative to the sun and natural atmospheric and geologic variations.

Climate Change: Severe weather caused by natural events and air pollution that insulates earth's surface.

How earth's energy is used to generate energy and power.

How and why energy and power create air pollution.

How and why air pollution affects unnatural climate change.

Who and where are the greatest air polluters on earth.

Methods to generate less polluting power and needed goods.

Vehicles: How they work, energy cost, and why they all pollute.

Air Pollution difference between electric cars v. gas cars.

Conservation of Energy: How to conserve and save money.

CONTENTS

Introduction

The context of this story is how earth was formed and has been polluted during a period of rapid population growth since the Industrial Revolution in 1770, when the need was to mass produce essential goods for human survival. There has always been a growing need for food, energy, and power, and it continues to grow. Because of careless use of fossil-fuel (coal and crude oil) energy on earth, today the need is to develop substantial alternative renewable energy forms and to address the growing problems of air pollution, extreme weather anomalies, and waste disposal around the globe.

Pollution exists everywhere on earth: on the land, in surface air, and the atmosphere, and in rivers, lakes, and oceans. It is a massive global problem, and one that is curable only by joint determination and discipline of every country and person. At the present pace of population growth, rising power usage, and uninspired pollution mitigation actions, temperature extremes will continue to worsen as the population is expected to grow from 7.8 billion today to 10 billion by the year 2050.

When this book was written, the world was slowly emerging from the COVID-19 pandemic, reduction of crude oil pumping in the United States, and the Russian war with its neighbor, Ukraine. Worldwide supply problems, inflation, global unrest, a rapid upsurge in air pollution, and abrupt climate change followed. The confluence of these disturbing factors is leading to tempestuous weather, perhaps a climate crisis if not combated.

Back in 1890, the population of earth was 1.53 billion, and the total energy generated was one terawatt, or 1,000 billion watts. One hundred years later in 1990 there were 5.52 billion humans on earth consuming 13.5 terawatts of energy, which is about four times as much energy used per person. The consequences of rapid energy demand are surface waste and air pollution throughout the globe.

In 2021, greenhouse emissions rose 6%, a rate faster than the economy with record heat in the Pacific NW and severe ice storms in Texas, followed two months later by opposite record temperatures in both places. By mid-year 2022, weather changes inferred that most regions on earth will begin experiencing weather they have never seen before, including torrential rainfall and flooding in Eastern Australia and Pakistan, and a near Category 5 Hurricane Ian in Florida. Pollution in one place on earth causes weather problems half-way around the globe in an unfair distribution of pollution and grief for those who do not pollute.

On May 20th of 2022, a region in Colorado experienced a snowfall just one day after a record 90°F temperature day. The global climate has shifted, rotating from south to north, and vice versa below the equator in unpredictable long term atmospheric circulations as both poles begin to warm.

In addition to weather extremes in India, a grain exporter, and drought elsewhere, the Russian War on Ukraine worsened the looming worldwide shortage of grain. This coupled with supply chain problems from the COVID pandemic led to accelerated inflation and acute food scarcity for up to 50 million people in Africa.

The second objective of this book is to provide readers with brief definitions of words, terms, and expressions used explaining the subject matter. It is not necessary to memorize these. The list is intended for introduction and as a reference.

A rigorous examination of the physics and chemistry of the subjects is complex. This book exams these in a simplified manner using everyday language without complex equations and chemical formulations. Skipping details in some chapters will not detract from the primary focus, combating climate change. Readers who wish to delve more deeply into attributes of specific topics are suggested to visit: *your topic- Wikipedia,* an acclaimed scientific encyclopedia.

I used Wikipedia for verification of some factual data, five photos, and I am a financial supporter of the free-use scientific encyclopedia.

> *What the world does not need is to go faster.*
> — Mahatma Gandhi

> It needs world leadership to combat climate change.
> —Patrick Simmons

*Note: The portion of Chapter 3 on the origin of the universe and our galaxy is discussed only to the extent that it applies to how we arrived at the present weather, and how future energy sources may aid in the battle to control pollution. It is **not** intended as a rigorous scientific or religious viewpoint of creation.*

CHAPTER 1

DARK CLOUDS OF POLLUTION WITNESSES

T his book is about the fundamentals of our galaxy, earth, our sources of energy, and how we convert energy into power to perform useful and necessary tasks in our lives and at what cost. The purpose of this book is to explain climate, weather, pollution, energy, and what we have done in the past and should be doing now to combat climate change.

The book assumes that the *beginning,* as far as pollution is concerned, was 250 years ago. This was the start of the Industrial Revolution around 1770, when humanity began using earth's resources in a more engineered, or *applied physics,* manner to mass produce goods, power, and food for a growing population.

It is important for readers to understand that earth was created and evolved with natural resources of potential energy such as water, wind, coal, crude oil, magnetism, thermal vents, and all the chemicals on earth, in the air, and under the ground. It is also important to know that earth is part of a solar system in which the sun and our atmosphere are sources of energy, but that earth is being polluted by our past ill-conceived highly consequential energy practices. The surface of earth in this story means the land above and below sea level, oceans and waterways, the atmosphere, and all life within the biosphere.

Readers are reminded that when any form of energy is converted into a different form of energy, there is energy loss sometimes referred to as friction, conversion loss, and inefficiency. A practical example is using electricity, gas, or oil to heat our homes and buildings. It takes energy and work to acquire these forms of energy, and when used, energy is lost in their conversion to heat. Conversely, the same applies to cooling buildings and conveyances in summer. Energy is never lost but made partially unusable.

Because of energy loss in the form of air pollution most places on earth will continue to gyrate between intense unseasonable warming and frigid cooling. Pollution causes ocean currents and the polar jet streams that control climate to alter from their natural variations to unpredictable shifts and subsequent tempestuous weather change.

This work is not intended as a historical document, nor is it a political missive. The story does not dwell into the economics of the various government administrative proposals. Comparative cost analysis of the various transportation and lifestyle alternatives are presented but not argued. These are topics of substantial debate, and time is the final judgement of the decisions made regarding various energy transformations proposed for the coming years. This book deals with the energy and power availabilities, usage options, and their impacts on pollution that affects all forms of life on earth, especially human life.

There are contentious political opinions about both the short term and long-term consequences of certain immediate and future steps to moderate *weather anomalies*. Some weather experts are not enthusiasts of the expression, *climate change crisis*, because they view the prevailing weather warming and climate variations as functions of human-caused pollution, not the natural historic, geologic, and atmospheric circulation pattern alterations. Air pollution is a major factor influencing air circulations and ocean currents which determine global climate and local weather.

Weather change, so-called the climate change crisis, is the result of misuse of fossil-fuel energy by the wealthiest countries on earth, not because of natural climatic variations. Poorer countries have not flourished as well because coal and crude oil did not improve their wealth or standard of living. Unfortunately, weather and pollution are global matters, not a single country's problem.

Artic and Antarctic temperatures are warming three to four times faster than the global average temperatures causing permafrost up to 500 feet thick covering about 10% of earth to melt. As surface ice melts, mineral laden water causes cracks that release methane into the atmosphere. This occurrence is affecting the weather on the rest of the globe.

Warming in the Artic causes more warming in a closed loop process much like audio feedback. In short, Artic warming begets more Artic warming. It might become irreversible with dire worldwide consequences. Antarctica is also starting to rapidly warm which complicates the climate and weather below the equator.

Global warming is not a disaster, there are many humans that freeze to death every year. Weather warming will spare thousands of lives from a painful demise, while others succumb to the opposite extreme. Planet earth has never been warmer than it is now. Five out of 10 of the last hurricanes to reach the United States were the costliest since 1990. Hurricane Ian in Florida at the end of September 2022 was the greatest destructive force since Katrina devastated Louisiana in year 2005. They are the result of coal and crude oil overuse without pollution control.

In Eastern Australia, including Sydney, historic flooding activity has caused loss of life, property damage, and severe commercial disruptions. Regions on both sides of the equator now experience the opposite, drought in the western United States, China, and north-eastern India. Drought causes the need for water rationing and reduced electrical power from hydroelectric dams. Unusual rainfall causes flooding, reduced food yield, and loss of life. Canada is battling historic forest fires across the country in June 2023.

The evidence suggests that air pollution and weather warming and cooling extremes will worsen during the next thirty years or longer as the population grows from 7.8 billion to 10 billion in 2050, and as the solutions are stifled by the whims of politics, arrogance, culture, and cupidity. Pollution around the world today causes five million deaths annually. Many of these souls have never seen the sun, moon, or stars while on earth because of air pollution, and they now see them as they rest atop the dark clouds of pollution victims as ghostly witnesses.

After reading this book if the reader becomes more aware of where energy is available, how energy generates power on earth, and the *environmental consequences* of various combinations of where and how energy is utilized or wasted, the effort will be worthwhile. The most effective solution to the impending climate change is not necessarily high technology nor electrifying everything, but rather education, common sense, and better energy use efficiency by everyone.

> *Environmental pollution is an incurable*
> *disease. It can only be prevented.*
> — Barry Commoner

CHAPTER 2

DEFINITION OF WORDS AND TERMS

The following words, phrases, and their meaning are used in this book:

Accrete, to enlarge by accumulation or coalescence.

Adiabatic means a closed process or system in which heat does not escape.

Adsorption, the process in which molecules or ions adhere to a surface rather than be absorbed by air away from it.

Air, a mixture of gaseous chemicals on earth that we breathe. Surface air is 79% N2 and 21% O2, with numerous air pollutants.

Ammonia, NH3, a natural molecule commonly used as fertilizer and as a source for extracting hydrogen, H2.

Analog, similar, related to, compared with.

Anode, the positive electrode of an electric process. In an EVS battery the anode is usually a graphite/nickel mixture that varies with manufacturers.

Argon **D**ating, a way to calculate ancient age using radiometry.

Asteroid, one of three types of solid material, C, M, and S that has existed in space since time began.

Atmosphere, air above earth, the sky up to 6,200 miles.

Atom, the smallest kind of ordinary matter forming a chemical element without an electrical charge.

Autonomous, self-controlled, not managed by external influence. In an electric vehicle it means autopilot mode.

Battery, a mixture of chemicals with a cathode, anode, and electrolyte able to produce and store electrical energy.

Big Bang, a non-scientific name given to the creation of our universe and earth's solar system.

Biosphere, the living (bio) creatures and plant life on earth and in the oceans.

Boyles Law, $P1V1 = P2V2$, relationship between pressure and volume when either is changed while temperature stays constant in a thermodynamic process.

Brine, water impregnated with salt, in the case of lithium mining, the weak alkaline is lithium hydroxide, $Li \cdot OH$.

Calorie, a unit of energy, $1 \text{ Cal} = 3.09$ ft-lb or 4.2 J

Carbon, C, is a unique element with four electrons enabling bounding to form many molecules. Diamonds and graphite are the purest solid carbon forms. CO and CO_2 are gaseous carbon forms.

Carbon Capture, a process of seizing CO_2 before it is released into the air, or a means of capturing CO_2 in the atmosphere. Also known as sequestering carbon.

Carbon Cycle, the biochemical cycle exchange of the carbon in the atmosphere of earth's biosphere.

Carbon Dioxide, CO_2, is a semi-toxic molecule of carbon and oxygen that pollutes the atmosphere, and that plants and trees convert to oxygen in sunlight.

Carbon Footprint, the sum of carbon released by any power usage where carbon pollution is involved.

Carnot Effect, n' the theoretical efficiency in percentage that a thermodynamic engine achieves when converting heat to work, $n' = w/Qh$, w is work and Qh is heat energy.

Catalyst, a chemical element that stimulates a reaction in a chemical process without it being consumed.

Cation, a positively charged ion that is attracted to the negative electrode in a battery or in an ion exchange process.

Cathode, the negative electrode in an electric process. In EVS batteries, a mixture of lithium, cobalt, and manganese is typical but varies by manufacturer.

Charles Law, $V1/T1 = V2/T2$, relationship between volume and temperature when either is changed, and pressure stays constant in a thermodynamic process.

Chemicals, one of about 120 elements natural to earth.

Chicxulub, the name of the asteroid that formed the Gulf of Mexico and buried the dinosaurs and extensive species of animals and plants on earth.

Climate, the position of earth's axis relative to the sun, and because of natural geologic and atmospheric influences, absent air pollution. Climate is a global phenomenon. Weather is always a local condition.

Climate Change Crisis, dangerous political misnomer, earth is polluted causing weather change, or *apparent* climate change.

Chlorophyll, green molecules in plants and some algae. Part of Its function is to aid absorption of photons and convert light energy to plant photosystems.

Coal, a black combustible carbon mineral formed from decayed organic life under pressure for millions of years.

Cobalt, Co, a magnetic chemical element mixed with others to form the cathode in electric vehicle batteries.

Conservation of Energy, the law saying energy cannot be created nor destroyed, only exchanged from one form to another with a portion made unusable.

Convection, heat transfer that occurs after the sun heats earth when surface air rises, and colder air is displaces it.

Crude Oil and *petroleum* are the thick natural hydrocarbon material found on and under earth and pumped from wells.

Cumulonimbus, the largest, highest cloud formations that with high pressure generate the energy released in lightning storms.

Cupidity, a word derived from French cupidite', excessive desire for power, control, and wealth, used by 1800's Montana industrialists.

Efficiency, the ratio of energy out to energy in, $n\% = Eout/Ein$. The ratio is always less than 100% in energy exchanges.

Electron, the smallest part of an atom with a negative electrical charge of -1eV. It is tiny compared to a neutron.

Energy, the ability to perform work. There are two basic kinds, potential and kinetic:

> a. Potential, energy because of position or storage: lake, river, dam, water tower, sun, lightning, fossil-fuels, wind, battery.

　　b. **K**inetic, energy due to motion: move, wind, tides, lever, rotation. Potential water becomes kinetic when used.

Enthalpy, the sum of the internal energy plus pressure times volume in a thermodynamic system.

Entropy, word meaning disorder. In a thermodynamic system it refers to the energy unavailable in a process, system, or reaction, i.e., wasted, unusable, or lost energy.

Equinox, equal day light and darkness when the sun is directly over the equator and tides radically rise - twice a year.

EVS, abbreviation for electric vehicles, battery, hybrid, or hydrogen fuel-cell powered. In this book it is singular or plural.

Externality, the indirect carbon cost of one party's activity to an uninvolved third party that unfairly burdens the third party.

Fission, atoms splitting into parts in a chemical reaction that releases thermal energy and usually radioactivity.

Fixation, conversion of volatile or free molecules to stable compounds in synthetic and natural processes.

Flaring, the practice by oil producing and refining companies to burn unwanted natural gas rather than capturing, storing, and converting it to hydrogen.

Force, F, any effort that causes matter to move, including gravity, usually expressed as, foot-pound or meter-gram units.

Fossil-fuel, hydrocarbons: coal, oil, gases formed in earth by decaying plants and animals used to generate electric power, for transportation, and producing food and industrial goods.

Friction, abrasion, rubbing, or viscosity are forms of energy waste or loss in machines and processes.

Fuel Cell, in general a device or system which uses H_2 to generate power like a battery, or to capture CO_2 from the air or in a carbon capture process such as in a power plant.

Fusion, atoms combining in a thermochemical reaction.

Galaxy, a solar system in space held together by gravity toward its sun consisting of planets, stars, and solar matter.

Gas, a fundamental state of matter expressed as molecules because of perfect mobility such as H_2, He_2, O_2, and CO_2.

Giga, 1×10^9, one billion as in quantity and frequency, gigawatts, or gigahertz.

Gram, unit of weight, one-gram equals ~1/28th once.

Graphite, pure carbon form, C, stable electrical conductor used as anode in batteries and in pencils, brake linings, and in heat transfer.

Gravity, the force between objects or energy to celestial mass. On earth gravity is defined by free-fall: $g = 32.2$ ft/sec/sec in a vacuum.

Greenhouse Gas, or GHG, is the combination of water vapor, CO_2, CH_4, dust particles, freon and other gases in the atmosphere.

Grid, the public and private electrical power distribution system to homes, industry, and commerce.

Helium, He_2, the second element on earth, an inert vital element lighter than air. He_2 and O_2 fused to create iron, Fe.

Horsepower, a measure of foot-pounds of force per unit of time. One horsepower equals a force of 550 foot-pounds per second, or 33,000 foot-pounds per minute.

Hurricane, a cyclical low-pressure tropical storm with wind speed of 74 mph or greater. In Asia, they are called typhoons.

Hydrogen, the 1st element on earth from which others were formed. This combustible gas comprises 75% of all known matter.

ICE, any type of internal combustible engine.

IEA, the International Energy Association.

Impedance, resistance to current in an AC circuit because of inductive and capacitive reactance plus ohmic resistance.

Inert, inactive, or unable to move. In chemistry it refers to noble gas elements He, Ne, Ar, Kr, Xe, and Rn.

Ion, an atom, or molecule with a + or − charge wherein the number of electrons and protons are unequal as in a lithium-ion or other battery exchanges.

Infrared, light energy below the visible range, <380 nanometers.

Iron, Fe, a heavy element created by the fusion of He2 and O2. Steel is Fe with carbon and base metals.

Isotope, two or more atoms with the same chemistry but with different numbers of neutrons and atomic weight. Some are stable and others are radioactive.

Joule, J, a unit of energy or heat power. One joule - second equals one watt. (Pronounced jowl).

Kilowatt, 1000 watts,1kw (elect) 1w = 1volt times 1 ampere, when expressed as energy or power used: kwh

Lidar, a type of short-range radar object detection using light which is higher in frequency than radar.

Light, the visible portion of electromagnetic radiation spectrum at wavelengths from ~400 to 700 nm.

Lithium, Li, the 3rd element is used as the cathode in lithium batteries, mined as solid Li_3CO_2 on land and $Li \cdot OH$ in brine. Pure lithium reacts with vigor in H_2O.

Lithosphere, scientific expression for the surface of earth.

Mass, m, the weight of an object or energy divided by gravity, $m = w/g$.

Matter, anything with mass from neutrons and protons to pachyderms and planets.

Methane, CH_4, active natural gas emitted from earth, sea, and ruminants. CH_4 decomposes up to ten times faster than carbon dioxide in the ozone layer.

Methanol, CH_3OH, the simplest alcohol, is used in hydrogen fuel cells and other chemical processes and can be used to power aircraft jet engines.

Mile, either Statute, 5,280 feet or about 1.6 km, kilometer. or Nautical, 6,060 feet or 0.001 degree at the equator.

Molecule, the smallest particle of one or more atoms held in a chemical bond without an electrical charge, e.g., liquid H_2O, gases CO_2, O_2..., and solids, $NaCl$, Li_3CO_2....

Monomer, a molecule that can bind itself to form polymers.

Nanometer, a unit of size or period, $1\ nm = 1 \times 10^{-9}$

Nebula, a distinct formation of H_2, ionized gas, and star dust that circulates between stars and is referred to in astronomy as a star nursery.

Neutron, n, part of the nucleus of an atom with zero electric charge and that is slightly larger than a proton.

Newton, Isaac, an English scientist who devised the laws of motion and conservation of energy.

Nickel, Ni, a magnetic hard element used in EVS battery cathodes and as a key ingredient of stainless steel.

Nitrogen, N2, the most copious (79%), chemical gas in surface air. The 6th or 7th most abundant element.

Oxygen, O2, makes up 21% of surface air and is produced on earth by plant-life.

Ozone, O3, triatomic oxygen from 9 to 22 miles above earth, peaking at ~18 miles high in the United States. Ozone is UV photolyzed O2, protecting earth from ultraviolet radiation.

Permafrost, ice up to 500 feet thick beneath the artic surfaces that traps methane and various natural gases.

Photon, a massless, sub-atomic particle of light, both natural and synthetic, whose energy is proportional to frequency.

Photosynthesis, the process in which photons convert the CO2 in plants to sugar while releasing O2 to the atmosphere.

Pollution, various contaminants in the atmosphere, land, and water, both man-made and as naturally occurring emissions. CO2 in excess is the primary cause of extreme weather.

Potash, water soluble potassium salts, most commonly potassium carbonate K2CO3, an essential plant fertilizer mined mainly in Canada, Russia, and China.

Power, the rate of doing work usually horsepower or kwh, power is work divided by time, P=w/t, 1 horsepower = 746 watts.

Proton, p, the subatomic part of an atom with a charge of +1ev that forms the nucleus with a neutron.

Q, symbol for heat energy, expressed as Joule, or one watt-second.

Rate, a quantity of something compared to a different quantity, as in feet per second or Kilowatts per hour.

Sequester, word used in carbon sequestering as a synonym for absorb, capture, remove…

Solar Equinox, two annual earth events when the sun is perpendicular to the equator, Vernal for spring in March and Automanual occurring in September.

Sorbent, materials with physical structures that can absorb or adsorb gases and liquids useful in carbon capture systems.

SPR, the strategic petroleum reserves in the U.S.A. where oil is stored in salt caves underground for emergency use or in wars.

Stratosphere, the atmospheric region from about 9 to 30 miles above sea level. It varies in height by latitude and pollution level.

Surface Air is generally meant to be surface to 6 miles or about 34,000 feet high.

Terawatt, a measure of energy, 1,000 billion watts, 1×10^{12}

Theia, the name of the massive asteroid event that scraped earth's surface creating the moon.

Total Energy Cost, the sum of carbon costs from each step in converting an energy source to performing work.

Troposphere, the atmospheric region from 5 to 10 miles high, with variations between the poles and equator.

Ultraviolet, UV light period above the visible range, beyond 740 nanometers.

Universe. The boundless sum of all energy, time, space, and matter from subatomic particles to galaxies.

Watt, unit of energy, 1-volt times 1-ampere.

Weather, what you see out the window or feel outdoors. Local atmospheric conditions at any place on earth are due to the position of the sun as influenced by natural events and human-made air pollution.

Weight, the quantity of mass exerted to overcome gravity, or the measure of force needed to move an object expressed as: w = mg, mass times gravity.

Work, applied energy, the transfer of energy from one form to another, or the relocation of energy and objects to another place.

CHAPTER 3

A BEGINNING: EARTH AND ENERGY

I n the beginning, if there was one, there would be subatomic particles and electro- magnetic fields (we think), then hydrogen, (we believe), then a light (we are pretty sure), at least a spark or flash to ignite the so- called *Big Bang*. This is the popular version of the beginning of earth and our solar system, but it does not explain the universe in which earth resides.

A theoretical explanation of *our* universe is that it was created by exploding a densely packed, near infinite quantity of hydrogen in emptiness. No astronomist, geophysicist, or earth scientist has ever clearly explained how the first molecule of hydrogen was created or how zigatons of H2 arrived in the middle of nowhere. Space is defined in physics as a boundless continuum in time, a definition just short of the word *infinite,* or beyond comprehension. This implies that the universe always was and always be.

There have been many models of the universe based on astronomical research, space probes, and advances in technical measurement instrumentation. These are theoretical and complex. The *universe* is better defined as the total of all energy, space, matter, and time, including all the gigantic galaxies and tiny sub-atomic particles, and everything we haven't discovered or understood.

Understanding the relationships between humans, earth, and the universe may be simpler using factorial notation arithmetic. N! means multiplying any number times every numeral before it, example: 4! = 4 x 3 x 2 x 1 = 24. 10! is 3,628,800, and 100! is a number that would fill three lines of his page. If you were in a city of 3.5 million people, earth would be (~10!), and our galaxy would be about 17 billion times (100!) compared to you or me. It is not possible to calculate the size of the universe.

Figure 3.1. Earth's Galaxy, the Milky Way

Today only about 5% of *our* universe and its energy is understood by Astro-scientists. New optics technology suggests the universe is probably infinite since previously observed stars routinely reappear as galaxies. Earth is in the galaxy called the *Milky Way* that consists of earth, the sun, planets, and billions of stars and solar dust.

The sun is a massive ball of burning hydrogen 93 million miles from earth and is the center of gravity of earth's galaxy. It is estimated to last just five billion more years. If you are at the right place devoid of artificial light at the right time the Milky Way might appear like Figure 3.1. The Milky Way was not the first galaxy in the universe. The order of existence is meaningless except for some astronomers and theoretical physicists.

Billions of years after the explosive sun event, cryptic clues were discovered that led humanity to an inkling as to what happened. Scientific   theories always begin with either something, energy, a force, or an assumption. From where did the first electron, neutron, and proton come? There are many theories, but no one on earth will ever know the answer to how or if it had a start. It simply is.

The Big Bang that occurred about 14 billion years ago, is the theory that at the time, concentrated hydrogen ignited, without postulating the origin of hydrogen or ignition source. The tremendous heat excited various super-active chemical materials, mostly hydrogen and then helium, which scattered and collided (fused together) forming all other chemical elements except iron (Fe). These elements formed the planetary building blocks in the circular system called a *solar nebula*.

Two hundred million years after the Big Bang, the sun was created, and all regional matter rotated around it. Later in geologic time, as matter coalesced, millions of stars and planets began to form, including earth. All chemical elements, except iron, were derived from hydrogen, the simplest atomic structure with one neutron, proton, and electron shown in Figure 3.2, the Bohr Model of the Atoms.

A new chemical element is formed when atoms are fused together because of energy from the Big Bang explosion, or from the extreme heat of other stellar explosions. The largest of these star systems are known as *supernovas,* including our expansive solar system.

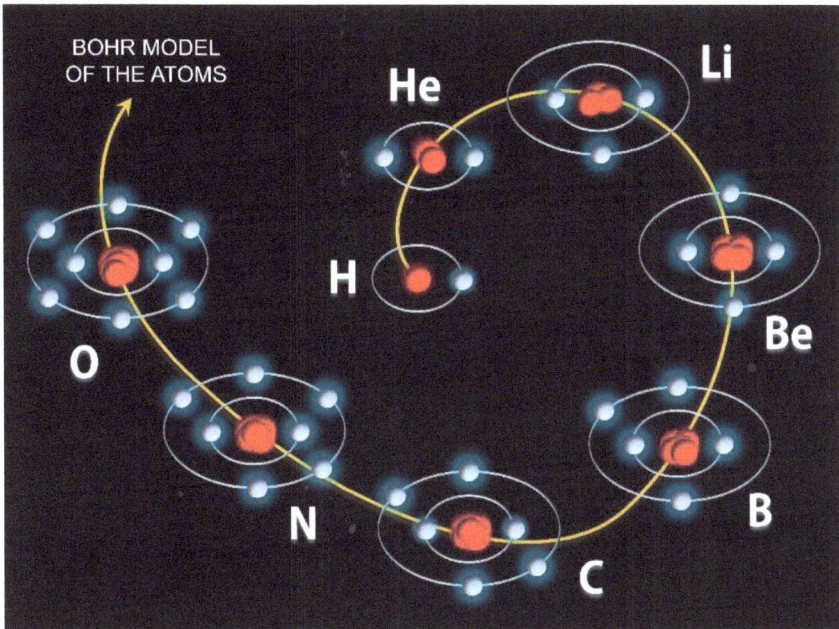

Figure 3.2 How the Chemical Elements Were Formed

A new chemical element is formed when atoms are fused together because of energy from the Big Bang explosion, or from the extreme heat of other stellar explosions. The largest of these star systems are known as *supernovas,* including our expansive solar system.

The Bohr Model of the Atoms depicts the fusion of two hydrogen (H) atoms to form helium (He) which then fuses with H2 to create lithium (Li). Hydrogen with only one electron is referred to as H2, because all gases become molecules due to their volatility that binds them together. In each progression, the new atom becomes heavier with more neutrons and protons forming the atom's nucleus. All atoms have an atomic weight that is the sum of the protons, neutrons, and electrons in the atom. The electrons of each atom circulate in valence rings and are extremely light weight compared with the nucleus. The element subscript shows the number of electrons in each element. The weight of gases is the same as the weight as the same elements in solid form.

Each valence ring consists of a finite maximum number of electrons starting with two, eight, 18, 18, and 32 in a mathematical arrangement as Atomic Numbers. The number of electrons in the outer ring decides the valence of an element. For example, the valence of H_2 is +1, it is willing to give up its electron. It is easy to see from the Bohr Model that two hydrogen atoms or an H_2 molecule plus an atom of oxygen combine to form water, H_2O. However, while it is easy to separate H_2 and O_2 in a high school chemistry lab, it is not possible to recombine the two gases back into water. That only happens in nature, or under uncommon, exceptional conditions.

The valences of N_2 and O_2 are -3 and -2 because they want electrons to fill their outer ring to eight. Carbon is complex, the C valence is usually + or - 4, it will either absorb or yield electrons depending on the circumstances. Carbon in CO_2 is a gas, and C in cubic form is graphite, or in its purest form is a diamond in which C atoms bind with themselves under extreme pressure. Carbon is everywhere on land and in the air, but it comprises only 0.025% of the chemical elements on earth. It is massive in the universe, the 4th most common element after gases H_2, O_2, and He_2.

All the chemical elements are arranged in the Periodic Table of the Elements by valence and other physical characteristics. The table is shown in color at the back of this book as Appendix II.

Fusion of hydrogen is the primary source of chemical elements on earth except for iron, which is unique, the only element non- fused from hydrogen. Iron was created as the result of helium and oxygen becoming fused in exploding stars and supernovas.

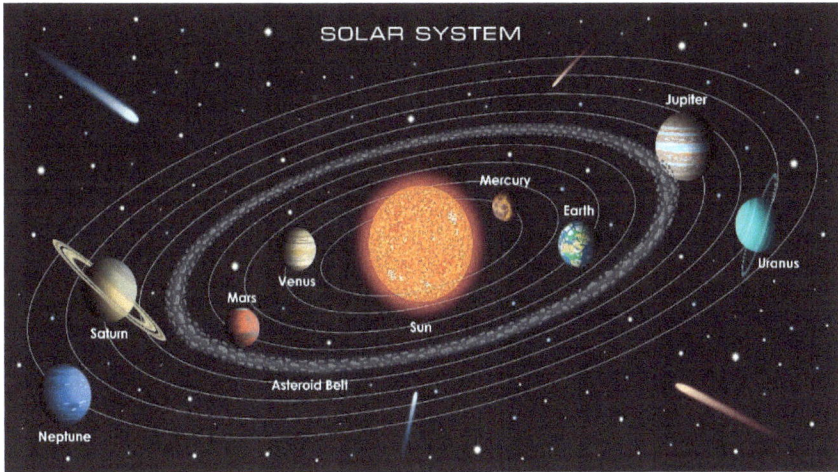

Figure 3.3 Formation of Earths Solar System

Energy from Big Bang powered rotation around the sun for billions of years until earth and its planets were formed as vast featureless hot planets. In geologic time, earth ceased to grow about 4.5 billion years ago, and its surface, called the lithosphere, began to cool. Hot magma caused by collisions of solar nebula remained active below the surface. Still deeper in earth, the core began to freeze and pushed outward. The resting place of earth settled about 93 million miles from the sun, believed to be the source of all energy in the galaxy.

Meanwhile, in the universe, galaxies like the Milky Way continued to appear and be filled with solar nebula, collection of material by gravity (accretion) occurred, and the planets and stars were formed in a manner like earth's creation. The solar system was filled with three types of asteroids of various sizes mainly in the asteroid belt, a region between the orbits of Mars and Jupiter. These minor planets vary in structure from stony clay S-types to carbon-rich C-type, and nickel-iron, M-types, the heaviest form. Asteroids are common in space and are found around earth in various sizes.

Two gigantic M asteroids had major impacts on earth creating the moon and the Gulf of Mexico. The largest asteroid remaining in space is called *Ceres* with a 600-mile diameter mass and is known as a dwarf planet.

In the first rare event in space, the only natural satellite of earth, the moon, was formed. The theory is that a giant object, an asteroid the size of Mars, travelling at about 56,000 miles per hour struck the side of earth in a gigantic glancing collision. The force, called the *Theia-Event* ripped the surface of earth into a giant rotating mass with its own center of gravity as it continued its journey into space. The gravity of earth determined the moon's final resting place 225,000 miles away rising and falling to viewers on earth every 28 days in phases from full reflection of the sun to near total darkness in solar eclipses.

The second major rare space event formed the geological conditions that over millennia stocked earth with over a trillion barrels of crude oil and other hydrocarbons, known as the *Chicxulub-event*. Impact of the gigantic collision between earth and an asteroid was the most significant factor altering earth's surface. It provided a massive energy resource whose careless overuse invigorated industries while unknowingly polluting the air. Chapters 5 and 6 explain the event that relates to the main climate topic throughout this book.

Another minor rare event in space happened in 1969 with the Apollo Moon Walk. Astronaut Neil Armstrong collected rock samples from the moon's surface and returned them to earth. Scientists analyzed these rocks, with later acquired samples, to calculate their age. The age was calculated by carbon dating, and it showed that moon is 85 million years younger than earth.

This was the time that earth's core settled with heavier minerals like iron and nickel declining beneath the surface leaving silica-based marble and conglomerate granite minerals on and near the surface. Volcanoes started erupting about a billion years ago because of earth's surface fissures. Magma was pushed upward by enormous gaseous pressure depositing nutrients on earth's surface. Water vapor and carbon dioxide were released into the atmosphere. Eventually clouds formed, and it began to rain, and it rained, until the earth was covered in water. Later still, internal earth pressure forced the 16 tectonic plates to rise and transpose until the seven continents we see globes were formed.

In addition to the volcanos seen on earth's surface, scientists have estimated that about 30,000 volcanos lie semi-dormant under the oceans. All volcanos emit various kinds of gases and solid materials including magma. At present 71% of the earth is covered with water, but only 2.5% of it is fresh water which will be affected by climate change and weather warming becoming scarcer.

Natural geologic and atmospheric phenomenon resulted in multiple earth climate changes to the extent that the South Pole was once the tropics, so was the North Pole a tropical zone at one time. There have always been *historic natural* atmospheric and geologic climate changes on earth evolving new forms of life.

When and how *homo sapiens* separated from the ape family is not clearly known. It took millions of years before humanoids migrated out of Africa to Europe and eventually crossed into Asia. They left Africa to find an environment with more sustainable resources as their population grew. How Australia was populated is widely debated because earth's tectonic plate positions and climate were much different than they are now. There is archeological evidence that the path was from the north via New Guinea. Other findings suggest crossings were from the south via Tasmania at about the same time, 60,000 years ago.

Figure 3.4 Neanderthals, About 50,000 Years Ago
Photo: Le Moustier by Charles R. Knight, Wikipedia

Human species underwent numerous variations millions of years before the early Neanderthals evolved. It was a biological and environmental necessity for humans to disperse around the globe to discover new life resources. A human tooth argon-dated 1.3 million years ago was found in Georgia, a European country northeast of Turkey, that times early emigration of humans out of Africa. Early fossils found in India suggest that humanoids may have existed 2.6 million years ago, but it wasn't until 1974 that a near-complete human skeleton, dubbed *Lucy* by the rockstars, *The Beatles,* was found in Ethiopia by Dr. Donald Johansen. The assembled skeleton is located at the Cleveland Museum of Natural History in Ohio.

Lucy was 3.6 feet tall and bipetal; her brain was the size of a Chimpanzee's; and she was argon-dated 3.2 million years ago. What is clear to genetic science is that humans evolved from a scientific black male and female form in Africa. Every human on earth has evidence in their DNA meaning all of us came from the same branch of the transformational tree.

When humans acquired a soul is unknown. Opinions vary, but ~200,000 years ago was when humans began expressing a belief in a higher power and religion in art, clothing, interactions, and earth symbols. Human intelligence, or IQ , is a gift from God. It is not related to race or wealth; it is a function of population.

Now earth is covered with man-made pollution and waste resulting in *apparent climate change*. The natural and historic climate changes extremely slowly, at a geologic pace. Climate is primarily a function of the earth's axis relative to the sun, atmospheric circulation, ocean currents, and geological activity. Humans are simply polluting earth into man-made weather extremes affecting all life.

> *Come together, right now, over* [climate].
> — The Beatles

Chapter 4

GLOBAL CLIMATE STATUS

T he reason for climate change is the natural reaction to air pollution caused by burning fossil-fuels: coal, crude oil, natural gases, and their derivatives that form molecular insulation around the globe. Without air pollution, climate change occurs slowly, and is the result of a natural event, continued tectonic plate shifting motions since early earth formation. Even minor earth surface shifts result in volcanoes and earthquakes on the surface and under the oceans that release carbon dioxide and methane.

Forest fires, lighting storms, blizzards, and cyclic atmospheric variations have existed since the biosphere began forming. Then as humans evolved using the natural resources of earth, minor amounts of air pollution began, and the rate has now become exponential resulting in rapid apparent climate change around the world. No country is immune from climate change because the atmosphere circulates constantly. It is a world matter, and every country needs to be part of combating it.

4.1 Global CO2

To combat climate change there is a continuous need to determine and monitor where and how air pollution is produced, who and what is responsible for it, and what measures must be taken now and, in the future, to combat it. Climate experts found that atmospheric carbon dioxide is the principal air pollutant with minor gases, and they measured the concentration of gases around earth. These molecules have been named greenhouse gases, GHG. They have been measured by geographic region in tons of CO2, percentage of world total of GHG, and emissions per person.

The purpose of this chapter is to present an overall perspective of the world climate change problem. It is not now a crisis, but without addressing the root causes and developing alternatives to how we generate and use power, life will become more uncomfortable and complicated. Climate change can increase drought, lower the fresh water supply, reduce crop yield, increase forest fires, cause inflation, stifle the economy, and affect the physical, mental, and emotional health of humans, animals, and plants throughout the world. Climate change is global, weather is local. These are the statistics:

The percentage of primary polluting molecules of GHG are:

CO2- 74%, CH4- 17.3%, N2O- 6.3%, others- %2.4.

The major GHG polluters by region and percentage are:

China- 27%, U.S.A.- 12.5%, India- 7.1%, Europe- 7.0%, and 46.4% the rest of the world that is thankfully widely scattered with many islands but contributing to the overall climate problem.

The amount of CO2 emitted annually in tons per person is interesting: U.S.A.- 19.6t/p, China- 9.1t/p, and India- 2.5t/p. China and India are the world's greatest coal burning polluters while the United States is the greatest polluter per person. Worldwide, the average amount of CO2 emitted per person is 6.27tons. Tons are metric or 2,205 pounds per ton.

The world produces 36.8 billion tons of CO2 per year with energy consumption accounting for about 76% of that. The rest of it comes from an array of sources which will be examined in greater detail in this book.   For us to combat climate change, it is important that we understand more about it and how the future will depend on what we do and don't do now.

Actions taken to date to reduce air pollution, including the transition to electric vehicles have worsened air pollution, and are not moderating climate change. Evolving technologies that have the potential for renewable clean energy are explained as well as nuclear fission that has the potential for abundant worldwide, emissions-free power generation without mining an energy source.

4.2 World Crude Oil Reserves

Earth acquired a tremendous supply of crude oil as the result of a giant asteroid collision that formed the Gulf of México, buried the dinosaurs and plant matter, and after about 66 million years provided earth with crude oil and many gaseous hydrocarbons. The total oil reserves by country and their consumption rates are of concern in addressing climate change as a function of burning hydrocarbons. The total estimated marketable oil reserves in billions of barrels in 2022 were:

Venezuela 304bbl, Saudi Arabia 260bbl, Iran 208bbl, Canada 170bbl, Iraq 141bbl, and U.S.A. was number 11 with 48bbl. *Marketable* means the proven reserves obtainable to sell at the current world market price for a profit.

The oil consumption rate by country in millions of barrels per day in 2022 were estimated to be:

U.S.A. 19.6mbl, China 12.9mbl, India 4.4mbl, Japan 4.0mbl and Russia 3.6mbl. The largest oil consumers are not the major producers of crude oil products. Dividing the oil reserves by the consumption rate for the United States by the number of years the reserves will last equals:

Years = reserves/ consumption rate per day/365

$$N = 48 \times (1 \times 10^6)/ 19.6 (1 \times 10^6)/365$$
$$= 48,000/19.6/365$$
$$= 48/7.154$$
$$= 6.7 \text{ years}$$

The United States oil reserve is likely understated because it does not include oil stored in the Strategic Petroleum Reserve (SPR) located in underground salt caves, nor oil stranded under federal property, and several huge known offshore oil fields like the Gulf of Mexico oil where it all started. The reason is that whatever is there is not readily available and marketable at the world market price for a profit.

Oil companies in the United States also have relationships with friendly countries in which they discover oil fields and lease the fields from foreign governments. The recovered crude is transported to the United States, or sold where it is pumped, and a portion is sold to other countries depending upon the market conditions. The oil supply is stable and long-term oriented. There is no danger of the United States running out of crude oil or natural gases in the foreseeable future.

The world oil supply is massive, and the market is a balance between the Middle Eastern Oil Producing and Exporting Countries, OPEC, and the primary consumers in Asia, Europe, the United States, and others. We would run out of oil in about 55 years only if the 2022 consumption rate continued and no new marketable deposits were found.

4.3 Pollution Politics

Politics is a part of everything including combating climate change and air pollution. This book attempts to be apolitical and mentions past events for information. Tax credits for purchasing

EVS vehicles and CO2 credit/debits are mentioned as being relevant to the intended audience. My opinion concerning socio-political agendas and behavior is that division or separation and reclassification of individuals or groups by natural unique features, aspects, traits, and personality then qualifying or celebrating these as deviances from the norm is counterproductive to society and in combating climate change. Divide and conquer is a political control scheme and war strategy. Uniting and thriving together is a healthier and stronger problem solution on earth.

> ESG *is the Devil…* [of energy and climate]
>
> — Elon Musk

CHAPTER 5

ENERGY FORMS ON EARTH

E arth is the only planet in our solar system ideally situated for human population because of our age and relativity to the sun. In addition to 120 known chemical elements in the earth and on its surface, other sources of energy include the sun, thermal, wind, nuclear, and magnetic energy. The common names of the five energy forms are: Chemical, Mechanical, Thermal, Electrical, and Magnetic.

Energy is classified as either *Potential* or *Kinetic*. Potential means energy forms that are available because of position. Kinetic energy refers to mass in motion and is the generator of power which is the ability to perform work. Water is potential energy when unused, and kinetic when used in a windmill. Oil is potential energy underground and kinetic when burned in power generating plants to produce electricity or as fuel in vehicles. Both classes of energy is discussed in this chapter, and a section is devoted to each of these classes and forms of energy. Figure 5.1 depicts well-known energy forms on earth with elevated stored water as a major energy resource.

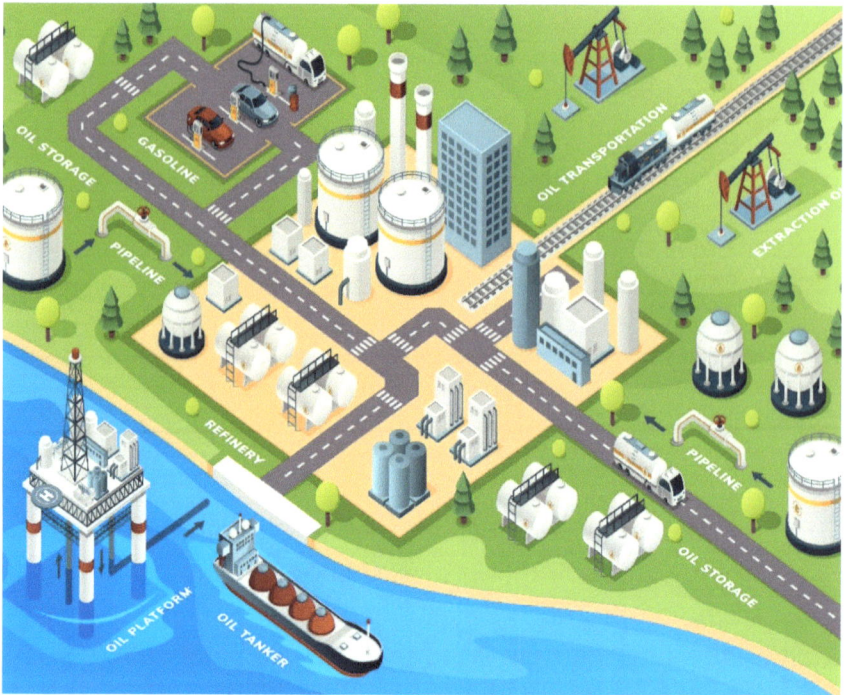

Figure 5.5 Common Energy Forms

5.1 Potential Energy

Water in a tower, or behind a dammed river or lake is potential energy converted to controllable kinetic energy. The potential energy behind a hydroelectric dam or stored in a tower is expressed as: U = mh, where U is potential energy in foot-pounds or in equivalent metric terms, m is mass or weight w divided by gravity, h is height in feet above a turbine generator, and u is surface loss. Common conversions are at the back of the book in Appendix I.

$$(1) \ U = (wh/g)-u$$

Gravity means weight or the pull of a large mass on objects above on its surface toward its center. On the surface of earth, it is expressed as acceleration, 32.2 feet/second/second. All objects, gas, solid or liquid feel the same force and will fall at the same rate regardless of weight unless they have aerodynamic characteristics. A pound of feathers will fall to earth at the same rate as a pound of sand if the feathers are compressed into the same shape as the bag of sand.

In static terms, water pressure increases with height at 2.3 pounds per square inch for each 12-inch height increase. For example, at a height of 17 feet, the pressure at the bottom end of a closed one square inch area pipe is:

$$(2) \; p = 2.3 \text{ lbs/sq.in} \times 17\text{ft} = 39.1 \text{ psi}$$

Fossil-fuel energy is abundant on earth in the forms of coal, crude oil, and natural gas. Most of these forms were created about 66 million years ago by the decomposition of plant and animal matter deep underground after earth was struck by a giant asteroid and is known as the *Chicxulub Event*. The asteroid that formed the Yucatan Gulf, commonly referred to in the United States as the Gulf of Mexico.

It was small, only ten-billionths the size of the Theia Moon Event, but it was a direct hit on earth, not a minor glancing scrape that created the moon. Its gigantic tsunami (wave) buried the dinosaurs, many other animal species, and much of the plant life on earth which in time became compressed. Over more time the decomposed material yielded fossil-fuels: coal, crude oil, and natural gases.

Oil was discovered in China hundreds of years before Christ (BC), but it was the discovery of oil in Pennsylvania by Colonel Drake in 1859 that led to the evolution of automobiles and oil-fired electric power plants. The first car powered by gasoline was patented to Mr. Carl Benz, a well-known German man, in 1886. The car's engine had less than one horsepower, but that quickly changed to much larger engines.

The Benz automobile became Mercedes Benz, followed shortly after by the designs of several others. Henry Ford introduced his crude quadracycle in 1896. The Ford Motor Company was founded in 1901 followed by Cadillac in 1902 at the upper end of the price scale. Ford's iconic Model A appeared in 1903 sporting the oval blue and white Ford symbol. Mass production of Model T Fords began in 1908 and continued through World War I until 1927 when the new Model A cars appeared on the market and became popular.

Figure 5.2 Henry Ford Quadracycle, 1896

World War I from 1914 until 1918 greatly expanded the engineering design and manufacturing capabilities in Asia, Europe, and the United States. The war effort accelerated mass production of improved machinery and engines for warcraft, including aircraft. The technology growth continued in the United States and Europe, and for several years after World War I. The world economy expanded until it collapsed in 1929 when the value of silver tanked.

The Great Depression of 1929 slowed car sales and mass production temporarily until 1933 when the 25th Amendment of the United States of American Constitution repealed the ban on alcohol and the gloomy unemployed celebrated. The economy began to slowly improve until 1941 when World War II forced car companies back to being war machine companies. Prior to that, Ford Motors was competing with many automobile companies in the United States, most notably Chrysler and General Motors. Cars were being mass produced in England, France, Germany, Italy, and Japan during the time.

Discovery of oil in Texas in 1901 slowly led to the Oil Gusher Age about 16 years before World War I. Crude oil was plentiful and inexpensive. Oil chemistry was studied, and efforts to improve processes in refinement quality were researched. Refineries were built, uses of raw oil other than fuel were invented, and crude oil became one of the most viable worldwide trading commodities. Crude oil could be made into plastics, roofing material, paint, cloth…an endless number of products worldwide. Demand and oil well discovery cycled upward, and the price rose from $15 per barrel in 1890 to over $125 for Brent Crude in 2022. A barrel of oil is 42 U.S. gallons or 168 quarts.

The unforeseen downside that would temper the ebullience over the wonderous miracle *black gold* was the over-use of fossil-fuel that resulted in air pollution and the present tempestuous weather warming and cooling. That is why there is a global effort to use alternate forms of energy to re-balance the natural carbon cycle.

This mission will require a massive transformation that cannot be made quickly, nor should it be. If fossil-fuel had been managed more thoughtfully, the current problem would be far less critical. The perceived endless supply of oil gave rise to ever increasing car horsepower as auto companies hyped their vehicle performance to an audience of enthusiastic, gullible buyers.

Gigantic industrial movements often create unexpected consequences and pollution and are best made in well-thought-out steps. The past missteps and the next major new power and transportation technology advances are examined in later chapters. Coal and crude oil burning for electric power resulted in earth now being warmer than it was three million years ago before humans evolved.

Figure 5.3 Early Days after the Oil Discovery

Readers are recommended to keep in mind that any time one form of energy is converted into a different form of energy, energy will apparently be lost. Energy cannot be created nor lost, only converted to a different form and unusable waste based on the Law of Conservation of Energy. All energy transformations or conversions are less than 100% efficient and some are *very inefficient*. For example, a 1930's hydraulic ram used to pump water is 5 to 9% efficient. Conversely, a direct current voltage exchange from a silicon solar cell to a higher electrical AC or DC voltage can be in the range of 90 to 96% efficient. Always consider efficiency in the choices of energy conversion, especially in times of a major energy transformation as we are in now.

5.2 Kinetic Energy

The second general form of energy is kinetic, which refers to motion. Moving forms of energy are well known, but the losses in some forms of energy conversion are less known. For example, if water is stored higher than where it is to be converted from potential to mechanical energy, and it flows through pipes, a conduit, or penstock, energy will be lost because of friction between the water and the conduit or pipe surfaces. This lost energy is called *viscosity*. It also refers to a lubricant's slippery quality.

Gasoline and oil fuel internal combustion engines, ICE, are the most common examples of kinetic energy where a piston and rod combination turn a rotating shaft that is coupled to circular or linear motion or to do something else. Automobile and truck tires move people and material. Ship and airplane engines turn turbine blades and propellers. Engines are primarily designed to create *force*, the ability to do work. The mathematical expression is simple in any conversion format, English or Metric.

(3) $F = ma = wv/gt$ where, $a = vf - vo /t$, and $m = w/g$

Force is in foot-pounds or meter-grams per unit time, mass is weight divided by gravity, $m = w/g$, a is velocity per second, and g is gravity = 32.2 feet or 9.815 meters per second per second. This formula is widely used in biology, chemistry, engineering, medicine, and physics. If any object is moving at velocity V1 and accelerates to velocity Vf, the force increases by 4 x Vf. It also implies that for every action there is a reaction as expressed in Isaac Newton's Laws of Motion and Conservation of Energy.

Note: English terms are used in this book because it is the most familiar to readers who have not studied science. Mathematical measurements can be converted to metrics by using Appendix I.

Figure 5.4 Speed Kills and Pollutes the Air

One horsepower, 1 HP, is the unit force needed to move 550 foot-pounds per second or 33,000 foot-pounds per minute, or equal to 746 watts of electric. The term originated from the measured work that a large horse could manage over a short period of time. If a human weightlifter can clean and jerk 275 pounds to seven feet in one second assuming velocity is constant during the lift, he has generated 0.11 HP instantaneous, but not for very long.

(4) HP = (275lb) (7 ft/1 sec) / (32.2 ft/sec²) (1 sec)/ 550
    = 275 (7) / (32.2) / 550
    = 59.78 / 550 = 0.109 HP

He has also lost more than that in caloric energy. The work capability of an average man is about 1/25th that of a horse short term. The equation, F = ma applies to everything that moves from planets to electrons, even neuron exchanges in the brain. The faster anything moves or accelerates the more energy is used and converted to waste, or unavailable energy. Any vehicle that accelerates quickly and often for no reason is wasting energy and natural resources, and worsening climate change by polluting the air. The differences between electric and gas cars will be explained.

There will always be energy lost in a rotating shaft, engines, motors, gears and so forth. All moving parts of any kind use energy and are less than 100% efficient in converting any form of energy to power and work. The simple application of the concept is seen in windmills, propellers, and in wind machines that convert wind energy into electrical current that powers the electric grids.

In the following subchapters, the basic forms of energy are described as kinetic energy with the theme being energy forms available to perform work.

5.2.1 Chemical

Earth is blessed with abundant amounts of chemical elements that can be used in the manufacture of batteries or be used in fossil-fuel or nuclear power plants. However, the amount of any chemical is finite. Because air pollution is damaging life on earth, there is a need to utilize alternate elements for power generation, like batteries for vehicles and silicon solar cells for homes and building roofs. There is also a limit to battery chemical elements.

The ideal known elements for large batteries are lithium (Li), nickel (Ni), and cobalt (Co). Ni and Co are both expensive and far less plentiful than Li. The dilemma is that the known amount of Li available for economic recovery is enough to make batteries for 1.0 billion cars and small trucks. The problem is that the existing ICE vehicles on earth to be replaced are about the same number.

In time, all the most useful chemical elements will become scarce and significantly more expensive as their usage increases unless new deposits are found. Large batteries with characteristics like lithium-batteries, except they are much heavier, are made from sodium (Na), a plentiful ingredient found in common table salt. They are not ideal for automotive use but are well suited for domestic and land-based energy storage including household solar cells and electrical power storage in gigantic solar farms. Salt is inexpensive.

The global atmosphere is less than 1% hydrogen, a combustible gas, which is now being used to power some vehicles. Hydrogen is expansive in our oceans but not infinite. Recovery of H2 from water is technically difficult and expensive. Highly active methane gas, CH4 produced under earth and by some creatures eventually decomposes in the atmosphere at a rate up to nine times faster than carbon dioxide, CO2, yielding H2 and carbon molecules. At present, most of the CH4 is burned or escapes into the atmosphere.

Figure 5.5 Natural Gas Shipping Depot

Of interest is that natural gas, which is 90% methane, can be converted to H2 and manageable fuel with fuel cells and various separation and recombination technologies. Vehicles, including aircraft, can burn H2 as a clean fuel source. Prototypes of hydrogen- powered aircraft have been designed, built, and tested. Commercial use models are being contemplated and are only a few years away from wide use, especially as air taxis to and from airports using batteries recharged with H2 fuel cells. Other aircraft fuels and biofuels are being developed. By the year 2050, aircraft will look and be powered differently than they are now.

The chemical minerals causing the most concern now are the fossilized materials: coal, crude oils, and gases. These molecules all have heavy concentrations of carbon that when burned emit CO2, CH4, carbon sulfides, nitrates, chlorides, and various minor pollutants. These fossil-fuel minerals are the *greatest polluters on earth* whose use requires carefully planned control and use reduction measures.

All foods are chemicals arranged in palatable combinations that promote human health and generate the energy to exist on earth and perform work. Food energy is expressed in calories. A calorie is the energy needed to raise the temperature of one gram of water one degree Centigrade. This definition means little to most people, one ounce is equal to about 28 grams and one degree Celsius is roughly 1.8 degrees Fahrenheit. The average ideal caloric intake for men and women is generally regarded as 2,500 and 2,000 calories per day. Age, lifestyle, health, body type, and profession are contributing factors which define the specific nutritional needs for everyone.

Nuclear energy is a form of chemical energy using enriched uranium U235 and other chemical elements in a nuclear fission (fuel splitting) sustained reaction. This subject and nuclear fusion (fuel combining) are discussed in detail in Chapter 6.

5.2.2 Mechanical

This form of energy is well known as engines, windmills, and anything mechanical or non-mechanical that moves. It is one of the earliest forms of work, manual hammering by chimpanzees and humans. It takes energy to open a nut, whether an acorn or a coconut. The lever and fulcrum were one other type of primitive energy combination to move a large mass. The Egyptian workers used various types of levers and pulleys during the construction of the pyramids and sphinxes. The workers converted food energy into manual labor which has been a form of energy conversion since humanity began.

Today, and since Henry Ford, there has been a rush to generate more horsepower from a mechanical engine for general use cars. It is mostly cultural and has expanded to a debatable sense as the weather rapidly warms and cools. The energy used, and is then wasted, to go from 0 to 60 mph in two or three seconds, appears to defy common sense. Yet, it is being designed into our electric vehicles today as a remedy for *climate change*. The more HP, the quicker the acceleration, and the more pollution. Electrifying everything is not the ultimate air pollution or weather warming solution. It is a step in the right direction for now. Manufactures must continue to build and market what people want.

Electric car (EVS) models in the 2023 Fall line-up consist mostly of luxury vehicles by Cadillac, Lucid, Fisker, Tesla, and Ford pick-up trucks and others priced from $70,000 to over $250,000. The average worldwide EVS price is over $69,000. The cost of battery minerals and semiconductor shortage are driving the prices. These vehicles are designed for sale to the wealthy and upper middle-class folks and do nothing to moderate air pollution. The reason is car manufacturers cannot make a profit now on low-cost EVS.

The United States is in an economic recession in 2023 and most people cannot afford a new EVS or many used cars that have become over-priced. It will take at least a decade before EVS begin to have a positive effect on air pollution. FedEx just announced in August 2022 that they are predicting the year 2040 as their goal to be using electric panel trucks and vans for parcel deliveries.

The world does not need 1,000+HP personal cars and pickup trucks for everyday short trips. Over 95% of trips in the United States can be made in a small low power vehicle according to the International Energy Commission. Many Asian countries use small cars and motorcycles for commuting. The higher the HP, the greater the carbon footprint. It is up to humankind to decide how we use and waste our energy, and whether we choose to pollute.

Pollution is often an unthoughtful choice made by habit or because auto companies have instilled an addictive cultural yearning called *hot-rod mentality* in the United States and

European countries. Rapid acceleration and speeding to a stop sign, then using hard braking, is a waste of energy and in most cases do not result in the driver reaching the destination sooner than conservative drivers. In addition, there has been a rapid rise in auto-related highway deaths in the years 2021 and 2022 despite fewer miles driven due to the COVD pandemic. Nearly on-half of the deaths were the result of drug or alcohol impaired drivers.

Chapters 8, 9, and 10 are devoted to automobiles, trucks, and transportation. These vehicles are well-known but are not well understood with respect to air pollution and climate.

5.2.3 Thermal

Thermal energy under the surface of earth exists and is being used in creative ways in countries like Iceland and Greenland, relatively cold places. In fact, Iceland has designed systems that use both a waste gas CO2 and thermal power to grow vegetables, even tropical fruit and generate electricity at the same time. Figure 5.6 depicts a greenhouse in a cold climate like that found in Iceland and Greenland wherein tropical vegetables and fruits are grown.

The way the system works is when a thermal vent is found and tapped, steam or hot water is delivered to a greenhouse complex where fruit and vegetables are planted. The input steam is condensed to water and heat in a turbine generator that produces electrical power and warmth for the trees and plants. The emitted CO2 in the air is partially captured and piped underground to nourish the plant roots. The plants and small trees absorb CO2 and release oxygen in a chemical reaction called *photosynthesis.*

Carbon dioxide is not an intrinsically evil gas, it is necessary for plant life. Plants from grass to large trees absorb and store CO2 in their roots; a wonderful way to moderate weather, lower pollution, save energy, and generate oxygen in the process. Thermal energy is extensive, is not technically difficult to harness, and is inexpensive to manage. Creative capture and use of thermal energy are warranted in regions around the globe wherever it is available and practical in the desired locations.

Figure 5.6 Thermal Greenhouse in Frigid Climate

5.2.4 Electrical

There are many ways to generate electrical power from potential energy including dams for hydro-electrical power; to kinetic energy in nuclear reactors, to solar cells for electricity, and in conversion of wind to areo-electrical energy. Electric conversion is one of the most economical methods to convert and use energy. The two common forms of electrical power are the alternating current of the AC power grid, and the direct current of batteries, generators, and solar cells. Elemental equations are used to explain these DC and AC circuits:

> **DC:** $V = IR$ known as Ohm's Law, where V or E is voltage in volts, I is current in amperes, and R is resistance in ohms. A volt-ohmmeter is used to measure electrical parameters. DC power: $P = IV = I^2R = V^2/R$ where P is watts.

Example: A 13.6-volt car battery powers up a 5-ohm spotlight so, V = 13.6 volts and R = 5 ohms.

$$13.6V \; o———/\/\/\/\/———o \; Ground$$

$$R = 5\Omega$$

(5) $I = V/R = 13.6/5 = 2.72$ amps
(6) $P = V^2/ R = (13.6 \times 13.6) /5 = 37$ watts

A simple system to understand with easy arithmetic. Please remember the relationship between power, voltage, and current, $P = IV$, to better understand electric vehicles (EVS) and batteries in later chapters. DC electric energy is produced by solar cells, auto generators, hydrogen fuel-cells and many types of batteries.

Most of the electric power generation on earth is in the AC power grid cycling at 60 Hertz (cycles) per second. Solid-state, high efficiency inverters are used to convert DC energy from solar cells and batteries to the AC form that can be directly coupled to the electric power grid or used in homes.

AC electrical power is primarily produced by rotating generators in dams and power plants that are fed into the power grid. The power grids are made up of many generators all operating at the same 60Hz frequency that energize the electrical power lines of the distribution system. This creates massive *inertia* which means the ability to remain at a constant frequency and voltage.

> **AC:** $V = IZ$ where Z is *impedance,* the combining of resistance with reactance XL and Xc to determine the AC current and power in the demonstration circuit as shown:

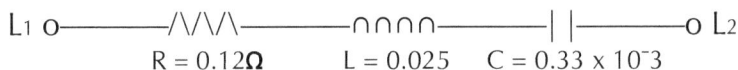

L₁ o————/\/\/\————nnnn————| |————o L₂
　　　　　R = 0.12Ω　　　L = 0.025　C = 0.33 x 10⁻3

(8) $Z = \frac{1}{\sqrt{R^2 + X^2}}$ where R is resistance and X is total reactance,

X = XL + Xc. Z is defined in ohms, the symbol is Ω.

Inductive reactance, $XL = 2\pi fL$, and Xc, capacitive reactance, $Xc = 1/2\pi fC$, where $\pi = 3.1416$, f is frequency in Hertz (cycles per unit time), L is in Henries, and C is in Farads. When Z is calculated, power is determined by $P = IV = $ watts, the same as with DC power. Reactance and impedance are expressed in ohms. If the voltage across L1 - L2 is 220 volts, 60 Hz.:

(9)　　XL = 2 (3.1416) (60) (0.025)
　　　　　= 0.09423 ohms

(10)　　Xc = 1/ 2(3.1416) (60) (0.33) (0.001)
　　　　　= 0.0 482 ohms

(11)　　X　= XL + Xc
　　　　　= 0.0942 + 0.0482
　　　　　= 0.049 ohms

(12)　　Z　= $1/\sqrt{(0.12)(0.12) + (0.049)(0.049)}$
　　　　　= $1/\sqrt{0.0144 + 0.0024}$
　　　　　= 1/ 0.0168 = 59.52 ohms

(13)   I   = V/ R

           = 220/59.52

           = 3.69 amperes

(14)   P   = IV

           = (.220) (3.69)

           = 812 watts

(15)   HP= P/ 746

           = 812/ 746

           = 1.09 HP

These equations are intended to demonstrate that advanced mathematics is not needed to acquire a basic understanding of AC electric circuits. The circuit shown is not real in practice. Electric motors and generators all have resistance and both capacitive and inductive reactance. Lights bulbs are resistance.

Electrical power generation does not occur in nature, except for lightning, it began with galvanic cells. In the 1820-30's when British scientist Michael Faraday experimented with coils of wire looped around a magnet and he detected electricity. His discovery and other developments led to the inventions of AC and DC motors and generators that became perfected. Eventually the tungsten light bulb was invented by Thomas Edison who was a promoter of DC electrical power generation, and the inventor who was awarded 512 United States and European patents. His DC circuits had major power distribution problems.

Edison hired a bright young engineer, Nikola Tesla, born in Serbia, and who immigrated to the United States, and they joined in 1884. Both men were scientific geniuses: Edison, the great thinker-inventor, and Tesla, the great calculator, soon began the great debate over AC and DC power systems. They parted ways after working together for only six months. The conflict was over which was the best method of electrifying cities, AC or DC power distribution systems. Both were stubbornly unrelenting in their respective beliefs and positions.

Tesla recognized the inherent inefficiency of high current DC power generation distribution and began work on transformers and alternating current generators. Tesla invented an array of AC motors, generators, transformers, and polyphase distribution technologies which he licensed to Westinghouse Electric in 1888. This provided him with financial funding and recognition as the key player in electrical engineering and the power distribution arena. Tesla also designed X-ray equipment and was a leader in radio engineering and related technologies.

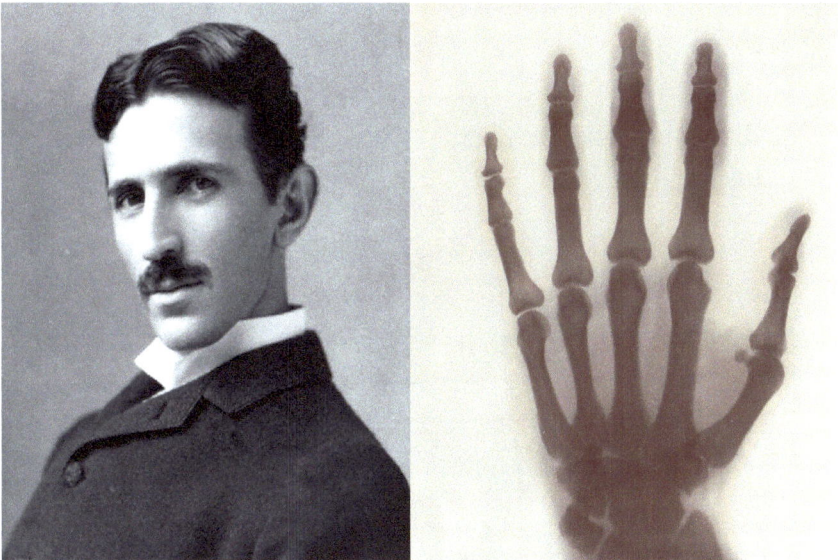

Figure 5.7 Nikola Tesla and X-ray of his Own Left Hand

Both Edison and Tesla received hundreds of patents, awards, grants, and accolades, but neither man ever won the Nobel Prize, despite having led the way to today's modern electrical power distribution systems around the world. The Noble Committee voted that the men applied known scientific facts and did not discover or create original science. Methods of electrical power generation are examined in detail in the next chapter.

5.2.5 Sunlight

The world has become aware of how to acquire and store energy from the sun by using photo-voltaic conversion from light directly into voltage, and by capturing the sun's energy to heat water. The sun is earth's primary source of energy because earth evolved from the sun. The sun emits photons that are used by plants to convert CO2 into O2 and produce sugar, or to heat water, and charge solar cell batteries. The word *photon* is Greek for light.

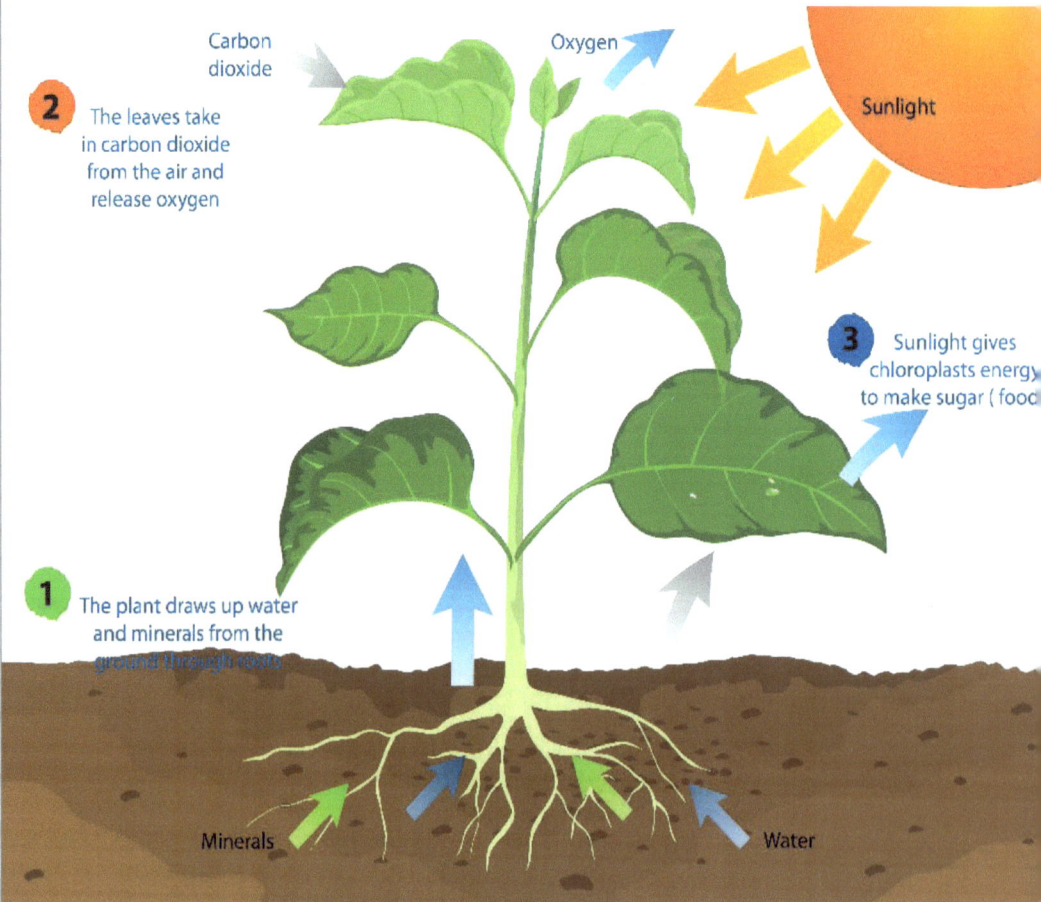

Carbon dioxide

Oxygen

Sunlight

2 The leaves take in carbon dioxide from the air and release oxygen

3 Sunlight gives chloroplasts energy to make sugar (food

1 The plant draws up water and minerals from the ground through roots

Minerals

Water

Figure 5.8 Photosynthesis Carbon Capture

All plants need CO2 for growth and the sun's photosynthesis to generate O2 and produce sugar for bats, birds, and bees. All forms of life need O2, and so does every combustion process, as in all types of furnaces, fossil-fuel engines, and forest fires.

The chemical exchange in *photosynthesis* is straightforward with understandable chemistry.

Sunlight, CO2 + Water yields Sugar + Oxygen

(16) 6CO2 + 6H2O ——> C6H12O6 + 6O2

In chemistry, a balanced formulation is calculated by counting the number of atoms on both sides of the exchange. Example, count the number of atoms of a molecule using the number of molecules times the valence number and add:

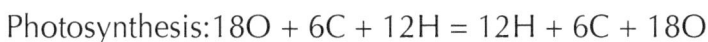

Photosynthesis:18O + 6C + 12H = 12H + 6C + 18O

A balanced chemical transformation.

Photons in sunlight trigger plant leaves and needles to excite chlorophyll, the green compound in plants, that when mixed with CO2 produces glucose, a basic sugar that feeds bats, birds, bees, and butterflies; and while feeding, they pollinate the plants on earth. Magnesium is a mineral in soil that initiates photosynthesis in a complex biochemical process.

Sunlight is the most economical energy form to generate electrical power and is essential for all plant life. Sunlight will last for five billion years, perhaps outliving life on earth. Humankind has just begun to harness the potential of the sun's energy, having been very late to utilize the primary source of all energy on earth. How this energy source was overlooked for so long is a mystery. Crude oil and coal help explain why humans did not discover and use it sooner.

Florida Power and Light (FPL) developed the world's largest solar battery in Parrish, Florida, in 2021. It consists of 132 battery containers each with their own inverter sufficient to power 329,000 homes for more than two hours. This amount of energy, available for power grid outages in hurricanes or huge lightning storms, is equivalent to 100 million Apple iPhone batteries, an interesting, applied energy comparison.

Earth has a limited amount of chemicals, but the sun will last for about five billion more years. Harnessing the sun's energy is a viable energy strategy to reduce CO_2, emit O_2, and in achieving carbon neutrality. Many small countries such as Netherlands, Luxemburg, and Singapore have grown vertically with creative architectural designs encouraging planting attractive small trees and plants throughout the sides of high-rise buildings and on the rooftops and balconies.

Architects and engineers in The Netherlands designed underground buildings with trees and bushes planted on the roofs. Photovoltaic cell arrays convert sun energy to electricity for internal use. Carbon dioxide in the building is piped to the roots of the plants growing on the roof in an efficient cost-effective closed- loop-system. These ingenious designs implement carbon capture, enhance pollution moderation, and reduce living costs for the citizenry.

Energy from the sun is not equally distributed on earth. It is a function of position in the seasons. The equinox occurs when the sun is perpendicular to the equator, meaning that sunrise and sunset are both equal. The northern limit is the vernal equinox in spring and the southern limit is the autumnal equinox when it is darkest in the United States. Sunlight is a major factor in farming, determining ideal planting and harvesting times. The equinox occurs twice each year and causes some of the most dynamic ocean tide extremes.

Sunlight has a direct influence both on earth and in the atmosphere where weather is structured. Later chapters in the book discuss pollutants, ozone, and the atmosphere, and how they interact with the sun and oceans affecting world climate.

5.2.6 Magnetic

Magnetic energy was defined by James Maxwell, a physicist/ mathematician who in 1861,62 expressed *light* as a frequency of electromagnetism in one of his equations on the subject. He showed that light beams behave the same as radio signals in radio engineering mathematics. Everyone is aware of small magnets and magnetic name plates on the sides of a refrigerator and on its door. In practice, the application of magnetic energy is much greater in our daily lives. Magnetic energy is used in generators, electric motors, and is widely used in electronics and information technology. Iron is the most common magnetic element, but nickel, cobalt, and gadolinium are also ferro-magnetic at room temperatures. Nickel and cobalt are important elements used in EVS lithium-ion batteries.

Earth is magnetized and magnetic fields encompass the globe. Normally, magnetic fields are invisible, but during periods of sunspot activity, they appear as spectacular arrays called the *Northern Lights.* They become visible near the artic circle in winter and at times shift downward to lower latitudes. These displays occur when electro-magnetic energy charges clouds in the upper portions of the atmosphere (ionosphere) with opposite polarities (+ and -). When sufficiently charged, water crystals and chemical elements in the cloud formations become ionized, and flashes of light appear across the full visible spectrum. The display is known as the Aurora Borealis in the north pole and the Aurora Australis in the south pole at opposite-season winters.

Figure 5.9 Aurora Borealis Over Basaltic Iceland

Lightning happens when a similar combination of electrostatic and electromagnetic energy occurs in the atmosphere. During storms, thunder cloud formations called *cumulonimbus clouds* form along a cold front, and they become charged with opposite polarities. These clouds can reach an altitude of 52,000 feet, but usually discharge at lower altitudes. Lightning starts after high pressure in the formation pushes water vapor upward. With rising altitude, ice crystals and soft hail called *graupel* are formed. Toward the top of the formation ice crystals become super-cooled and collide with graupel causing higher pressure. The cloud top breaks horizontally forming an *anvil shape* because of wind shear, and the voltage difference reaches a level sufficient to ionize molecules in the sky resulting in tremendously powerful electric flashes.

There are three types of lightning, -cloud to +cloud, -cloud to earth, and earth to +cloud. In a massive storm, all three types can happen nearly simultaneously. The energy released in lightning flashes is enormous. In February of 2022, a record flash 477 miles long from Texas to Mississippi was observed by satellites during a severe storm. The storm was so unusually massive and extremely high that 40 of Tesla CEO Elon Musk's SpaceX internet satellites were destroyed. These were very small replaceable satellites. On July 24, 2022, 53 satellites were launched by SpaceX to replace them and add a few to the formation.

Air pollution is a significant cause of increasingly violent storms in most parts of the earth. The scientific name for lightning is *disambiguation*, which to us means neutralizing the tension in highly charged cloud formations. For those who want to know more about lightning, the *Wikipedia* site has vivid graphic color illustrations set in motion of cumulonimbus cloud activity and lightning flashes. The displays are vivid and intriguing to watch.

Hurricanes are the most violent types of storms influenced by air pollution. They have happened for many centuries. The first recorded hurricane was seen and written into logs by Christopher Columbus in 1495. The air was not polluted then, but five of the 10 costliest hurricanes in the United States have occurred since 1990 because of air pollution. The two worst hurricanes in the Atlantic were Katrina hitting Louisiana in 2005 and Ian that made landfall near Fort Myers, Florida in 2022.

Tornadoes are the second most destructive storms during when violent spiral winds can lift homes and small buildings off their foundations and relocate trucks and cars. In the United States they frequently begin in southwest Texas and travel eastward across southern states or veer north crossing the central plains to the mid-Atlantic. The worst of these storms have thousands of lightning flashes, heavy rain or hail, and cause loss of life and destruction of property.

Over long periods of time, the polarity of magnetic earth has reversed. In the distant future what is now the magnetic north pole may become the south pole. The position of earth relative to the sun and as shown on maps and globes has nothing to do with magnetic polarity. Earth will not become upside-down for a long time from now. There have been several magnetic pole reversals on earth since time on earth began about 14.5 billion years ago.

Humanity has not used the full potential of magnetic energy due in part to the complexity of the subject, but energy is there for possible future use.

> If *you want to find the secrets of the universe, think in terms of energy, frequency, and vibration.*
> — *Nikola Tesla*

CHAPTER 6

APPLICATIONS OF ENERGY

I n Chapter 5 the sources and forms of energy were described. This chapter examines where energy is found and how it can be utilized to generate the electrical power needed for continued human development. Of primary concern is the extent that energy transformations emit pollutants and contribute to changing weather. Because of the vastness of this topic, it is separated into the most obvious or common energy sources, then into the more specialized forms needed and may be available for moderating the pollution that is impacting severe weather fluctuations.

Transportation is intentionally limited in this chapter because the subject is discussed in detail in Chapters 8, 9, and 10. Moving goods and people requires a massive amount of applied energy and is a major cause of air pollution.

6.1 Common Applied Energy Forms

Some of the most visible energy sources are storage of water behind river dams for hydro-electric power generation, fossil-fuel power plants, and mass transportation. Also, of interest and concern are coal mining for power plants, solar panels for electricity and heat, and wind turbine electric generators. Other sources, such as rare earths, are less known and scarce, but are becoming vital materials. Rare earth minerals are not actually rare. They are difficult to find, and gathering and processing them uses large amounts of land causing environmental concerns. About 90% of rare earths are mined in and exported from China. They have applications in military technology and in future development of air pollution mitigation technologies including permanent magnets for electric motors and generators, and in optics.

6.1.1 Hydroelectric Power

There are hundreds of mechanical and hydro-electric dams built on earth. Water wheels have been used for centuries BC to rotate a shaft and perform work. The work was mechanical: grinding, pounding, transferring water for human use, irrigating crops, and preventing soil erosions and floods. The Panama Canal is an example of mechanical work performed by large dams by controlling its waterway levels to move ships. Dams are an economical source of apparent free energy on land if there is a continuous supply of fresh water.

In some areas the supply of water stored is decreasing because of weather warming caused by air pollution that results in severe drought in places. Drought enables forest fires and causes rapid land evaporation. If the earth can recover from pollution, over- harvesting of trees and plants, and depletion of forests and jungles for other purposes, many reservoirs will fill, and the dams will flourish to generate renewable electrical power and controllable agricultural irrigation. It depends mainly on how we react to prevent more air pollution that has a 200-year head start.

Hydroelectric power generation in the United States began with a small generator energized by Niagara Falls, New York in 1881. Thomas Edison and Nikola Tesla participated in early hydroelectric power generation of both AC and DC electric power networks. Early-stage projects grew to about 41 hydroelectric power stations in the United States and Canada by 1886. Many of these dams were privately-owned plants related to the mining and logging industries. By 1890, there were about 200 hydroelectric small dams in the United States and Canada.

Figure 6.1 Hauser Dam near Helena, MT 1908

After World War I, 1914 to 1918, the United States, Canada, and many European countries fell into brief, but painful economic recessions followed by nearly a decade of recovery because of modernization of industries during the war. However, by 1929 highly overvalued stocks sold off, crashing the stock market in America. That triggered a ripple effect on foreign stock markets. Most parts of the world fell into a decade of economic and emotional depression. Banks around the world stopped lending.

The U.S. Treasury devalued precious metals, and other factors led to the Great Depression that lasted for 10 years until 1939.

As part of the United States government's recovery effort, the Bureau of Reclamation Plan hired thousands of people and started irrigation projects in the western states that led to the construction of Hoover Dam and Lake Mead south of Las Vegas, Nevada in 1928. The dam was on the boundary between Nevada and Arizona. Lake Mead became a popular destination for boaters, water skiers, and fishermen. The primary purpose of the dam was to supply electric power and water to California, not especially to Arizona and New Mexico.

Hoover was the world's largest hydroelectric power generator at 1,345 mW. The U. S. Corp of Engineers became involved in hydropower and flood control, and the corps completed Bonneville Dam on the Columbia River in 1937, followed upriver by Grand Coulee Dam in 1942. Dozens of other dams were being constructed on large rivers and lakes in the United States and Canada during the same period. Hydroelectric dams are marvelous ways to generate electrical power if the level of water behind the dam remains reliable.

Because of air pollution, forest fires, and drought the water level behind Hoover Dam in late July 2022 was the lowest ever following its peak level. This was followed in August 2022 by the Department of Interior decreasing by 20% the amount of Colorado River available to vegetable farms in AZ and CA, and by reducing by 8% the amount of water diverted to NV and NM.

Figure 6.2 Lake Mead Behind Hoover Dam

Worldwide there was an ever-increasing demand for electrical power. There was also a wicked hunger for socio-political power that led to aggressive conduct by a dictator in Europe and an emperor in Asia during the late 1930's. World War II was in its beginning stages. Electrical power generation became more urgent during and after the six-year war fought on the European and Asian battle grounds after the Japanese attack on Pearl Harbor, Hawaii in 1941. Reconstruction placed renewed stress on electric power grids around the world and the power demand has grown ever since.

Renewable energy hydroelectric dams were being designed and constructed around the world after WWII. The Itaipu Dam in Brazil became the world's largest hydropower dam in 1984 at 14gW. It was surpassed by the controversial gigantic 22.5gW Three Gorges Dam in central China in year 2008. By year 2020, hydroelectric power accounted for 37% of the renewable power in the United States, and about 17% of the global supply of electrical power.

Dams of all types, designs, and purposes have advantages and downsides as in other forms and uses of energy. The advantages of hydroelectric power are efficiency and low-cost electrical power to industry and housing, flood control, near zero emissions of air pollutants, ability to manage power output level, and dams are a renewable energy source. They also supply a reservoir for biological enhancement and recreational uses for nearby residents. More importantly, hydraulic power generation has no impact on weather change or the global climate cycle.

The disadvantages are lesser known. Hydroelectric dams displace people and reduce the surface land available for other purposes, like homes, agriculture, and forests. In the case of The Three Gorges in China, its construction displaced 1.24 million residents, killed 33,000 people, and forced 18 million citizens to relocate to higher or distant ground. In addition to the construction costs of dams; the electric generators, switching station, and power lines are expensive compared with other types of power generation. Their construction also requires extensive energy use by massive earth moving and building machinery that emit enormous air pollutants and greenhouse gases from diesel engine exhaust.

The other downsides are the drought problem and questionable safety records. In the last 60 years there have been periods of severe drought in various parts of the world during which hydro power output was either reduced or ceased. Lake Mead is now at the lowest level in its history as the power demand in California continues to rise.

Figure 6.3 Failed Hauser Dam on Missouri River 1909
Photo by unknown author, Magazine of Western History

Occasionally, dams fail because of unusually heavy winter snowpack followed by an early snow melt, torrential rainfall, typhons, or engineering design flaws. The failure of large dams has resulted in thousands of deaths around the world and in about 100 or so in the United States.

The seventh and last *Territorial* Governor of Montana, Samuel T. Hauser, an acclaimed civil engineer-industrialist, designed and built several hydroelectric dams for copper and silver smelters. His largest dam, an iconic steel design on the Missouri River, one of only three steel dams in the world, failed in 1909 due to an early heavy snow melt that undercut the dam footings flooding much of Helena, MT. No one was injured, but property was damaged, and the claims were personally paid by Mr. Hauser who rebuilt the dam in 1911.

Samuel T. Hauser was the author's great-grandfather. He worked as an engineer until age 81 and passed away at age 84 in 1914, with a lost fortune about when WWI began.

6.1.2 Fossil-Fuels: Coal, Crude Oil, and Gas

Crude oil, coal, and natural gas are abundant on earth, but they are far from an infinite energy source. The use of crude oil is important today not only for vehicle fuel refinement for all moving vehicles, trains, and ships, but for many other industries. Plastics, building materials, solvents, paint, rubber tires, medical supplies, laboratory devices and other materials and objects are made from crude oil. Crude oil will remain a vital raw material irrespective of other alternative energy sources. The use of coal as a fossil-fuel is the worst air polluter on earth and should be reduced and eliminated as soon as possible.

The impulsive curtailment of United States crude oil production in 2021 was unwarranted. That step worsened inflation, disrupted the world oil market, and did nothing toward moderating air pollution. It was a misguided political whim apparently designed to solve the impending *climate change crisis* by forcing a premeditated transition to electric vehicles. Electrifying everything is not a practical solution to moderating weather extremes or solving air pollution. The obvious questions are from where will the growing electrical recharging power demand come? From where will the needed minerals be sourced? Contentious opinions abound!

The reasons that severe weather and global warming will continue are that alternative power sources require new factories to be constructed; increased mining and transportation of heavy materials will occur; and electrical power generation demand and computer usage will continue to increase. These are air pollution producers, not solutions. A more prudent action would have been the encouragement of lower-power, general-use fossil-fuel vehicles with a gradual, industry-led transition to EVS. American car manufacturers were not prepared to jump into EVS auto mass production. This is why American EVS are so expensive. Ford and GM admit that they cannot make a profit on low-cost EVS cars.

Then the *lock-down-everything* response to the COVID-19 pandemic compounded the 2021-22 rate of inflation in the United States by closing schools and paying working people not to work, as well as severely disrupting the global supply chain. This step resulted in the worst inflation in 40 years, a 50% stock market decline, a shortage of microchips for technical products like EVS, and a general scarcity of commonly used products, medicines, baby formula, paper, and many varieties of food.

As the world population grows, small gas-powered cars with over 60 miles per gallon using improved catalytic converters for low harmful emissions may emerge, especially for use in large cities in the United States, and more so in lesser developed countries. The reason is economics. Inflation caused the increase in price of all cars, products, and services, especially oil and gas, and the rapid rise in cost of a new electric vehicles designed and advertised with high horsepower and fast acceleration. These EVS did nothing to moderate air pollution, they aggravated the problem because only the wealthy upper class could afford them.

Powering vehicles with methane is 30% less polluting than using gasoline. Methane, CH4, in the form of natural gas, is also recovered from oil wells and is in use around the world for vehicle fuel today. Powering vehicles with methane is 30% less polluting than using gasoline. The problem with fossil-fuels is the emission of carbon dioxide and carbon monoxide. The most polluting man-made objects on earth are electric power plants, manufacturing, mining, cargo ships, large trucks, planes, and gas automobiles.

Power plants around the world burn coal, crude oil, and natural gas. Together these fuels account for 61% of the electrical power generation on earth. Of these, coal is by far the worst fuel, but its use produces 37% of the world's electrical energy. Fossil-fuels emit most of the CO2, CO, nitrogen oxides, sulfides, and other flue gas contaminants in the atmosphere around the world. The collection and disposal of airborne and surface waste material is a major environmental problem for fossil-fuel power plants. All power plants of every kind use water ingress for cooling wherein the egress water becomes contaminated; and in some cases, it must be treated to avoid poisoning plants and animals. Fresh water per person becomes scarcer every day. causes

Coal is a sedimentary mineral made up of mainly of carbon with various combinations of H2, sulfur, O2, and N2. It belongs to a family of materials called *lignite* and is found in most places around the globe. It has been used for centuries BC as a fuel and was the vital fuel resource generating steam early in the Industrial Revolution.

Coal is the cheapest and most abundant fossil fuel on earth. In a coal-fired power plant, raw coal is first crushed to about two cubic inches plus dust and is blown into a furnace where it is converted from thermal energy to mechanical energy, and finally to electrical energy. Each complex step wastes energy creates surface heat, dust, and emits great greenhouse gas pollution in various forms. In addition, the mined coal must be transported from the mine to the power plant by train, barge, ship, and truck; all are major polluters. Conveyors are used to move the raw coal both to and from the methods of transport, another huge waste of energy and air pollution source.

In addition to its inefficiency, coal is neither a pure carbon nor a bound molecule. It creates substantial surface waste and water contamination. Coal contains many other minerals including arsenic, mercury, thorium, and uranium. The emissions of sulfides and nitrides in coal causes acid rain, especially in Asia and India, where 60% of the coal on earth is burned. The presence of uranium and its radioactive isotopes in coal-fired power plants exposes the plant operators and nearby residents to three times as much radiation as a nuclear power plant with the same electrical energy output.

Today, it is the primary cause of the air pollution on earth. The major coal mining countries are China, India, Australia, Indonesia, and Russia. Graphite is one of the hardest forms of coal, and is used as the anode in lithium batteries, and in powdered form as a lubricant. China is the primary source of graphite and vital materials used in electric vehicles.

Figure 6.3 Coal-Fired Electric Generating Plant

Huge fossil-fuel power plants around the world not only became major land surface and water polluters, their construction and operations deplete the forests that were growing near their sources of water. Forests are essentially the opposite of power plants. They absorb CO_2, release oxygen, and make the world a livable planet.

The effect of deforestation is even more pronounced in the tropical regions. Brazil was once the epicenter of CO_2 absorption, but logging, agricultural, and cattle ranching interests depleted millions of acres of highly CO_2 absorbing forests.

The elimination or conversion of coal fired power plants to natural gas or oil is the most urgent and effective way to moderate weather change. There are no known practical methods to reduce the harmful effects of using coal to generate electricity. Technology advancements have created ways to capture power plant CO_2 called *scrubbers,* whose use is effective on natural gas fired power plants, but they are useless on coal fired plants.

Carbon dioxide, as mentioned earlier, is not an intrinsically evil gas. Plants need CO_2 and sunlight to generate oxygen and sugar in photosynthesis. Certain types of scrubbers can capture CO_2 and store it underground to feed the roots of plants and trees. These iconic practices can be found on numerous power plant stacks in many non-Indo/Asian countries around the globe. These and other pollution moderating measures are discussed in detail in Chapter 9.

6.1.3 Mining

The most useful solid minerals on earth are found by mining in some format. Minerals are found in wells, veins, lodes, placers, deep pits, and in the oceans. The most abundant of these are iron ore, coal, and chemical elements to make metals, concrete, and fertilizers. Iron is mined on every continent except Antarctica, usually on surface mines. Copper and its chemical relatives are found in huge deep pits. Precious metals are usually found in mineral veins.

Lithium, a soft white metal, is the primary element mined for electric vehicle batteries. Lithium hydroxide, Li·OH, is found underground as brine. Solid Li is found on surface mines as the mineral lithium carbonate, Li2CO3. Many minerals are found as byproducts in the search for something more plentiful or valuable such as magnesium with lithium. In this book, the primary interest is not how to discover and mine ores, but what are the environmental consequences of mining.

Mining disturbs the surface, requires enormous energy, fresh water, and releases pollutants into the air and local water supply. Many mining operations use harsh chemicals, solvents, acids, and even mercury, to recover the desired mineral. There are other downsides.

Because of careless operatives and environmental damage, a conference called the Rio Earth Event, was organized in 1992 in Brazil, S. A., to map out a set of mining standards and regulations. Financial interests and politicians vetoed that approach and argued for self-regulation. Mining around the world has used child labor, unfair wages, minimal safety standards, and has caused environmental damage, sickness, and great loss of life.

Coal is found in large veins in rolling hills often with a surface outcropping revealing its location. Mining coal is usually in the form of tunnels into the sides of hills. Tunnels are dangerous places to work. Without proper bracing the loosely packed material atop a tunnel can cave in, and the air can be poisonous the deeper the tunnel. Early coal miners used a canary in coal mines as an air quality indicator. If the canary dies, get out or you may be the next to go!

Figure 6.4 Coal Miners Deep in Tunnel

Workers tend to be men unable to find better jobs, and the chores are burdensome in an unhealthy environment. Even more dehumanizing, coal mining takes thousands of lives every year by accident, cancer, or other lung diseases.

Every country has its own regulations, but in most of the major mining countries members now comply with and support the International Operational Standardization (IOS) principles. While it is difficult to measure, the air pollution emissions from mining releases huge amounts of CO_2 from diesel fuel; uses electrical energy; and releases toxins, dust, and fossil particulates into the air, ground, and water supplies. The countries with the greatest amounts of vital minerals pollute the most, but the atmosphere is in continuous random motion circulated by powerful jet streams. Pollution in any region affects the entire world and is the major cause of severe weather fluctuations we call climate change.

Cement is a vital mineral form that is a combination of mining plentiful limestone and crushing it into powder. It is the primary ingredient in concrete that holds the sand and aggregates together in the mixture. Cement is made by crushing and screening limestone that is mostly silicon and calcium, then heating it to 2700°F in huge rotating kilns. The kiln output is marble-size spheres called *clinkers* that are ground into fine powder usually called Portland Cement. The noisy, smokey process emits gases and fine dust into the air and ground. But cement is essential on earth and irreplaceable.

On a global basis, cement making, and concrete mixing are responsible for 5% of air pollution. Concrete and masonry cement are the most common products on earth. Concrete and steel are the foundation and mechanical structure of buildings and highways around the globe.

Not all mining is performed for the purpose of removing minerals from earth. Some are mined and then be refined and processed into objects for use on farms. *Fertilization* is a form of mining in which minerals are removed, then processed, moved, and returned to earth to nourish and promote agricultural products and landscaping. The three most used fertilizers are ammonia, nitrogen hydroxide $NH3$, and potash as potassium carbonate $K2CO3$ or as a chloride KCl, and several other K salts. The fourth most important plant nutrient is phosphorous, P.

In the spring of 2022, there was a severe shortage of potash because of the Russian War on Ukraine. The largest sources of potash are Russia, China, Belarus, and Saskatchewan Providence, Canada. The war disrupted supply chains, and fertilizer imports from Russia were sanctioned. The curtailment of the Keystone Pipeline in 2021 coupled with the war in Ukraine aggravated matters because the price of both crude oil and natural gas, which is the primary sources of ammonia, dramatically rose. The result for most people on earth was inflation, higher priced commodities, expensive fertilizer, and food insecurity.

Figure 6.5 Potash Mine in Northern Europe

Ammonia is a stable caustic, toxic gas with a pungent smell that occurs in nature as a gas, liquid, or solid. It is also found throughout the solar system and on planets. Ammonia bonds with other elements to form numerous compounds widely used in chemistry. It easily burns with oxygen and is used in engines on land and as rocket fuel. About 85 to 88% of ammonia is used as fertilizer in the form of *urea,* H2NCONH3, also called carbonic acid and aqueous ammonium.

Compounds of potassium, KCl, KSO4 and KNO3, are naturally occurring salts that are vital nutrients in all plants. They are mined from deposits deep underground, surface strip mines, or leached from brine deposits. Magnesium sulfate, MgSO4, is usually found combined with potassium sulfate, KSO4. Magnesium sulfate, sold in stores as Epsom Salts, is essential for the formation of chlorophyll in plants that react with sunlight in photosynthesis. The United States has enough magnesium for its agricultural needs without importing, but many other fertilizers are imported.

6.1.4 Forests and Agriculture

Forests and jungles are the natural carbon balancing mechanisms on earth and are a source of energy. Forestry is the science and practice of management of federal, state, and private forests. Trees are harvested for a wide variety of building materials, and the softwood tree cell fibers are treated to manufacture cardboard and paper products.

Forestry is a relatively new science that evolved to prevent depletion of some species, and to limit clear cutting of an entire area. In the western states of the United States, there have been debates as to the wisdom of selective harvesting versus clear cutting of timber and undergrowth. The decision is based primarily on environmental concerns and preservation of a species.

In the parts of the United States and Canada, trees have been over-harvested for lumber and building materials. Since plants began growing on earth, there have been forest fires that are healthy natural events that clear underbrush and fertilize trees.

Unusually frequent forest fires in the last few decades are partially the result of poor forest management by utility companies in California that resulted from excessive accumulations of ground fuel. Lightning or arson sparked a flame igniting surplus fuel and when homes were set afire it became uncontrollable. It is crucial to recover the burnt forests by replanting to reduce pollution, capture CO_2, generate oxygen, recover animal species, and building materials. Building homes in the middle of a forest is unwise.

Logging operations create air and land pollution and damage the environment. After trees are harvested for commercial usage, the ground should be restored, the trees replanted, and fertilizer applied to accelerate regrowth. Drought is naturally occurring, but weather warming air pollution insulates the globe and stifles normal balance. Its outcome leads to an unpreventable ill effect on forests around the world resulting in plant death and soil erosion, as well as extinction of animals, birds, and organisms.

The worst recent damage to forests by lightning has happened in Canada, the western United States and the temperate forests of Europe and central Africa. Spain and France had massive forest fires in 2022 destroying vineyards and agriculture. The British Columbia Forest fires in June 2023 darkened New York City and about a quarter of the countries skies as hundreds of fires trans versed the country. The United States measured 5.9 million acres of forest fires by early August of 2022, with hundreds of fires either partially or completely uncontrolled with two months left in the dangerous fire season. These were historic records not the usual.

Processing of trees has become a highly mechanized and computerized industry. Modern lumber mills use electronics and computer programs to make decisions on the cuts made on logs to yield the most demanded and highest valued board sizes. In plywood mills, scanning and automated technology is used to cut peeled wood strips with knot holes to maximize yield and minimize waste. Other systems detect sparks, smoke, and warn to prevent fires. In the United States, the industry has been responsive to energy conservation and clean air policy. That is not the case worldwide.

Brazil, some countries in Africa, and in other places, tropical regions have been decimated by burning, instead of harvesting the wood. This practice, motivated mainly by greed and haste releases tons of air polluting CO_2 and shrinks the natural carbon absorbing function of tropical forests and savannahs. There are few measures.

Figure 6.6 Deforestation of Forests and Jungles

other countries can take against the whims of politicians and wealthy landowners in another country.

From August 2019 to July of 2020, Brazil cut down and burned 4,280 square miles of rainforest, primarily in savannahs for agriculture and cattle grazing. This was 9.5% greater deforestation than the previous year, and the practice is continuing. The world is losing its greatest natural carbon capture and air purifying system. Replacing it will be difficult… more realistically impossible.

Agricultural practices use the energy on earth to energize human and animal populations. Nutrients, fertilizer, availability of suitable land, and sunlight evaluates crops in each crop producing region and determines whether there is a need for importing or exporting meats, foods, and products. Cattle belch methane that decomposes about nine times faster than CO_2, some minor CO_2, and returns chemical fertilizer to earth as manure. The human and mechanical energy applied to growing crops or raising and feeding cattle and ruminants is exchanged for food and the livelihoods of the inhabitants in each area.

The process tends to be efficient and clean except for grain harvesting that generates dust and burns diesel fuel, and for cattle raising which has become highly controversial and debatable. The combination of these two is thought to account for about 10% of air pollution in the United States. The types of air pollution are significant in the efforts to minimize their cause. Diesel oil burning machines till the soil; then plant and fertilize seeds; and finally harvests the crops but creates dust. Means to abate the exhaust and capture the dust are discussed in Chapter 8.

Regarding ruminants, the primary concern is cattle, dairy cows, and sheep, but not monogastric pigs. Ruminants are four-legged herbivores with a true stomach and three added digestive compartments. The digestive process uses saliva, bacteria, fungi, and protozoa in a fermentation process that generates methane which can be reduced by adding seaweed or some types of algae oils to the feed. Growing macro red algae does not require fresh water, land, or fertilizer; but in a highly competitive business, any added costs are unwelcome. Another possible solution to rudiment methane gas belching may come from re-engineering the DNA of fermenting bacteria in the digestive organs.

There have been politically motivated calls to eliminate cattle as a source of high protein foods. The protests hinge around the claim that cattle are significant generators of air pollution, and the pastures could be better used for solar cell arrays that reduce land available for human food. The fact is that cattle and dairy cows belch CH4 that decomposes quickly compared with their insignificant CO2 emissions.

Today there are about 31 million combined dairy cows and cattle in the United States. Toward the beginning of the 19th century, there were over 60 million American Buffalo across the country with no indication that they harmed either the land or the air quality.

During the 2020-22 COVID pandemic, plant-based or fake meat came to the market with claims that it was a viable solution to climate change and was healthier for human consumption. The primary source of protein in many fake meats is soybeans that are called the most complete plant protein containing all amino acids essential for human development. Soybeans are one of many legumes' native to Eastern China and have been used as Tofu centuries BC.

The *beyond meat* fake hamburger analog is more expensive than cattle meat or other animal protein ground meats. The taste? Taste is a chemoreceptor sensation that individuals can mentally change. In that sense, taste tends to become an acquired sensation. A person usually tends to choose and like or dislike any food based on pre-adult environment and parental influence. Humans will eat anything rather than starve to death, enjoy the flavor or not.

6.2 Specialized Energy

The transformation from overuse of crude oil to less polluting energy is emerging. The most obvious change is from fossil-fueled cars to EVS powered by lithium batteries. Lithium is used as the cathode in batteries blended with cobalt and nickel and other minerals. Graphite is the preferred anode. Battery companies continuously experiment with electrode materials based on price, availability, and performance to achieve their best battery.

6.2.1 Lithium Batteries

Lithium, Li a soft white metal, the third simplest chemical element, is abundant with one-half of earth's reserves found in Argentina, Bolivia, and Chile, South America. It is also found in several other countries both as a hard-rock form in the Democratic Republic of Congo, Africa; as surface mines in the United States and Canada; and as lithium hydroxide, $Li \cdot OH$, found in brine in parts of the United States and in many countries. There are also massive amounts of Li in oceans, but it is difficult and expensive to recover, and is unlikely to ever be a practical source for battery minerals.

The estimated number of lithium-reserves on earth is about 21 million tons, roughly 1.3 times the amount of tin, a heavy metal. Assumptions about the amount of Li needed for an EV car battery, studies have generally agreed that about one billion electric vehicle batteries could be produced. However, the reserves of Li are in various forms, and the energy needed to extract, purify, and render it into a chemically viable molecule widely differs. It remains to be seen what the average life of an EVS lithium-battery is, and how well they can be recycled without causing more air pollution than the lithium and other elements is worth.

There are three other materials in a battery that are not mined and refined in volume in the United States, cobalt, graphite, and nickel. Lithium extraction is not as simple as mining coal or iron ore. The current processes require extensive land and use a great amount of water, salts, and concentrated acids. The untreated wastewater creates substantial environmental problems which include water and ground contamination; unhealthy chemical reactions whose fumes affect humans and animals; as well as creating extensive air pollution. If lithium can be efficiently recovered from used batteries without emitting air pollution, it will be beneficial, but achieving that goal is doubtful.

Lithium mining companies that recover and process, Li·OH, from brine are desperately experimenting with an extraction process called Direct Lithium Extraction, or DLE, in which various types of membranes, filters, and electro-chemical exchange techniques that are used in place of concentrated acids to convert Li·OH to battery grade, 99.5% lithium carbonate Li2CO3 mineral powder form. While DLE is a promising goal, the objective has not been achieved as of mid-year 2023, at least it has not been officially announced with details.

Figure 6.7 Lithium Mining in Argentina

Another downside is the price of Li. It has risen from $12 per kilogram in April 2021 to over $80 per kilogram in March of 2022. Unless significant new Li reserves are found and DLE processes are improved, the price of Li will continue to rise as EVS production pace accelerates as expected. But as the supply of Li declines and the demand increases, the price rises except for the largest EVS manufacturer. Lithium batteries also pose a fire hazard in part because of chemistry, and due to the high voltage necessary for efficient EVS usage.

Lithium batteries are manufactured by connecting hundreds of small cells in series/parallel. The more cells in series, the higher the voltage; the more connected in parallel, the high the current and power. Each cell provides about 3.0-volt DC rising to 3.6 volts when fully charged. EVS are moved by AC motors of various designs, typically from 100 to 300 horsepower each, with one to four motors per car depending on performance design goals.

Chapter 5 said that one horsepower is equal to 746 watts. With a 200 HP AC motor at full power as an example, the motor energy input is more than 149,200 watts. The current capacity and cost of copper wire is proportional to the diameter or cross section area of the wire. To limit current to 200 amperes using the expression $P = IV$ and solving for $V = P/I$, the battery voltage must be at least 746 volts DC which an inverter changes to three-phase AC. This presents a potential high voltage problem for EVS, some of which have caught on fire because of battery flammability. Lithium in pure form vigorously reacts with water so care must be used in handling it during battery manufacturing.

Until the cost of EVS batteries declines, the price of an EVS is expected to significantly exceed that of a similar gasoline ICE powered automobile. Replacing the battery in an EVS is very expensive; and fast charging, now being promoted, decreases the life of Li batteries. EVS manufacturers dream of a million-mile battery, but nothing approaching even a thousand-mile battery exists as of mid-year 2023. Lithium will become scarce as the year 2030 is approached.

Less expensive batteries that use sodium from extensively mined common salt, NaCl, as the cathode have been built. They are also non-flammable. Their problems are heavy weight and less desirable charging characteristics. They have non-EVS uses in homes and buildings to store electric power from roof-top solar panels.

6.2.2 Nuclear Energy

The first nuclear chain reactor was developed at the University of Chicago Argonne Laboratories in 1942 by a team of physicists and engineers led by Nobel Laureate Enrico Fermi. In only 21 years later, the 1,150 mW Enrico Fermi Nuclear Power Plant on the shores of Lake Erie went critical in 1963. The boiling water (BWR) nuclear reactor was designed by General Electric's Atomic Power Division in San Jose, California. In the same year, the GE 500 mW N-reactor (NPR), the largest reactor ever built at the Hanford Atomic Works in Washington state went *critical* or became power active. This occurred about two months after President John F. Kennedy, who initially dedicated the N-reactor, was shot, and killed in Dallas, Texas, just weeks before it went critical in 1963. I was in the N-reactor control room at the time.

There was an important difference between the two nuclear reactors. The Fermi reactor generated electricity to energize the power grid. The Hanford reactors, designated A through M, were all designed to breed Plutonium, Pu239, primarily for nuclear weapons, and several other radio-active isotopes for medical uses. The N-reactor was designed to perform both functions. It was an electrical power generator and a Pu breeder. The problem with the early nuclear power reactors is that they were costly and took five to six years to design and build.

The Hanford Atomic Works was formed in 1943 as part of the Manhattan Project to breed Pu239 from U235 for use in the first atomic bombs against Japan in 1946. The N-reactor, or NPR for New Production Reactor, was the last of the Hanford nuclear reactors and it was designed as a combination reactor. N-reactor supplied electrical power to the Washington Public Power Service from 1963 through 1987. It was shut down and decommissioned in 1994.

General Electric Company was the leader in BWR's and was also assigned management and control of the Hanford Atomic Works in 1947 by the United States Atomic Energy Commission. However, around 1973, the nuclear reactor started to lose favor in the United States and most European countries because of highly publicized minor engineering failures in which no one was injured, and there was never environmental damage.

The story was different at Chernobyl, Russia where a nuclear reactor melted. In 2017, GE Hitachi Nuclear Energy was formally founded, although GE and Hitachi had worked in an alliance to build BWR nuclear power plants in Japan for about 60 years before 2017.

Figure 6.8 Hanford, WA, GE Nuclear Reactor N
Photo: United States Department of Energy

Westinghouse Electric Company developed a different approach to nuclear reactors by designing pressurized nuclear reactors (PNR) which were more suitable for powering large ships and submarines. The first U.S. Navy nuclear fuel vessel was a submarine, the USS Nautilus, launched in 1955, followed by the USS Long Beach, a heavy cruiser in 1961. By 2022, there were more than 160 nuclear war ships around the world, with 83 in the U.S. Navy fleet.

Nuclear energy has been generating electrical power on land and in ships since the mid-1940's. In a nuclear reactor, enriched uranium, U235, is split by neutrons into smaller nuclei in a chain reaction known as nuclear fission. The rate of the reaction, called the *nuclear period*, is monitored by multiple sensors in neutron flux monitoring instruments and logic safety circuits. The power rate is restrained by boron-infused control rods positioned between the uranium cruciform fuel bundles. The reaction generates heat which is removed by a cooling system that converts the heat to steam powering a turbine and electric generator connected to the power lines that form the electric distribution grid.

There are about 450 nuclear power reactors in the world accounting for 10% of total electrical energy generated on earth. The advantages of nuclear power include a large supply of uranium equal to the amount of tin on earth. The capital investment in a nuclear power plant is very high, but the operating costs are relatively low. More importantly now is the low carbon footprint. Nuclear energy is carbon-clean with low ground pollution and minor greenhouse gas emissions. Cooling water egress is significantly less polluting than the water used in equivalent power output fossil-fueled electric power generating plants.

Nuclear power technology has advanced remarkably from the first large scale, high-cost power plants to more flexible smaller nuclear reactors. The Department of Energy recently published an article describing five advanced nuclear demonstration reactors, one of which yields 15 mw and can be installed in only 30 days. Nuclear fuel pellets are now made using uranium-nitride for higher performance and longer life.

Other advances in nuclear technology include smaller reactors whose designs are more efficient and economical than older versions. Some of these designs are applicable for replacing the power systems on pollution generating fossil-fuel ships when weather changes become worse than they are now. Readers who are interested in new nuclear power developments can access the Department of Energy article by logging on to, www.energy.gov/ne/articles/ardw.

Nuclear power generation has undergone cycles of popularity in the United States because of highly publicized accidents and the implied poor safety that resulted in public fear. After initial praise of nuclear power in the mid-1940's until the early 1970's, a down cycle occurred for several years, followed by a ramp-up in nuclear power plants around the world. With the present state of earth's atmosphere and severe weather conditions, new nuclear power designs can contribute to moderating the climbing global temperature problems. Nuclear reactors are next to solar farms as the safest and cleanest forms of electric power generation. They also use much less land and do not depend on weather.

Nuclear fusion is another promising energy source that attempts to generate heat the way the sun does by smashing hydrogen atoms together. When H2 atoms collide at high speed, they fuse (combine) to form helium (He2), an inert gas. The process releases enormous energy in the form of heat equivalent to or exceeding the temperature of the sun. The reaction does not cause radioactivity or any type of pollution but H2 is not free.

As of January 2022, two experimental reactors, one in China, and one in the U.S.A., have achieved brief periods of nuclear fusion with the heat level exceeding the surface temperature of the sun which is ~6,000°C. In December 2022, scientists at the Lawrence Radiation Laboratory in CA were able to use lasers rather than linear accelerators to collide H2 atoms. That could simplify fusion, but not the extreme heat issue.

The greatest problem with nuclear fusion is how to contain the enormous energy and extremely elevated temperatures that occur during fusion. Development of a complex high temperature containment structure; thermal energy control and management technology; and energy delivery systems are decades away; and are technically challenging, but theoretically possible. Fusion does not require any fuel except H2 and emits no air pollution. Fusion electric power plants may become the answer to one of the greatest energy needs on earth if the temperature problems are solved.

6.2.3 Hydrogen Fuel

Hydrogen, H2 is plentiful, and a highly active element being considered for EVS, ships, and cargo trucks. Water is the primary source of H2 on earth. At present, the separation technology of H2 from water is expensive, impractical, and remains under research and development. It is more economical to use methane, CH4, in natural gas as the H2 source in fuel cells that can replace lithium batteries. There are also many advantages to burning H2 as a fuel because of virtually zero CO2, other harmful emissions, and an abundant fuel supply. The major disadvantages are that H2 is the most efficient at -212°C, which is near absolute zero temperature, and handling, transporting, and storing H2 is challenging, but becoming manageable and safer.

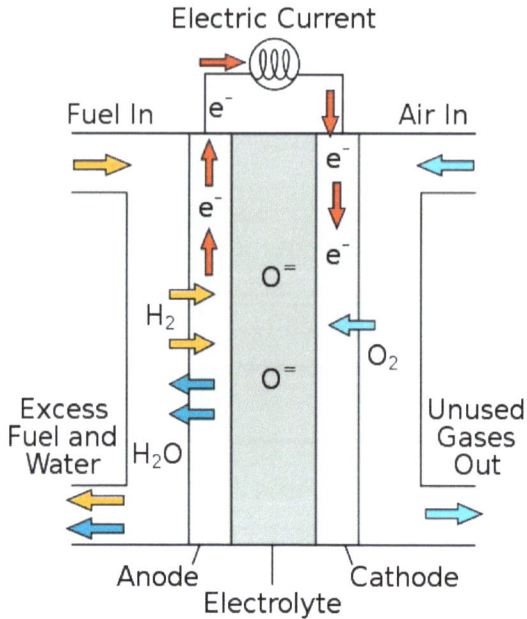

Figure 6.9 Hydrogen Fuel Cell Schematic Diagram
Photo: R. Dervisughi contribution to Wikipedia

Hydrogen fuel usage is currently limited to fuel cells which are designed to replace large batteries and as fuel in some types of aircraft. Figure 6.9 is a schematic of an individual cell that explains the electro-chemical process. Hydrogen and water is injected into the fuel cell anode chamber and air is routed to the cathode of the device that has an electrolytic core. The cathode is usually a Periodic Table Group 1A element like lithium mixed with cobalt and manganese that can exchange ions with a conductive element like graphite. The electrolyte serves the same purpose as in all batteries; namely, to provide a path for ion exchange, and is usually a solid or paste alkaline with potassium or sodium as the base mineral.

The ion exchange results in the flow of current from the anode to cathode that develops a voltage across the external load. To initialize the process, the anode is doped with powder platinum and the cathode with nickel. These are the catalysts which are not consumed in the ion exchange process. The fuel cell process is continuous once started if fuel and air are available. Fuel cells typically generate 0.6 to 0.7 volt per cell and are connected in series/parallel configurations in complete truck or marine *modules* to achieve the same performance characteristics as a lithium-ion battery. The theoretical efficiency of a fuel cell is about 50% which is twice that of a 25% efficient ICE gas engine. However, they are not as efficient as a lithium-ion battery and are awkward for small EVS.

Fuel cell technology has advanced mainly with designs suitable for mass transportation both as long-haul trucking EVS operations and for marine applications. The advantages are improved efficiency with larger-sized cells, long life, low weight, and hydrogen refueling does not draw from the power grid and takes only minutes. The disadvantages are the current cost of hydrogen and the initial cost of parts needed to store the hydrogen and control the process in vehicles. In time the disadvantages may be overcome.

Aircraft and ICE devices can use hydrogen as fuel, but safety concerns are limiting applications to research and development. Rolls-Royce, a British jet engine producer, believes that hydrogen powered aircraft will become a reality by the mid-2030's. When hydrogen technology becomes further developed, it will lead to wide-spread energy use in a cleaner air world.

6.2.4 Solar Energy

The sun is essentially a gigantic ball of burning hydrogen, emitting sub-atomic particles called photons that behave like radio signals. A photon is a massless particle continuously emitted from the sun. The word, *photon* is Greek for light. When photons impinge on a properly designed silicon photodiode surface, it releases electrical current that develops 0.5 to 0.62 volt across the junction of silicon and a dissimilar material, typically silicon fused with germanium or another chemical element.

Sunlight is captured on polyvinyl webs known as *solar cells* which can be placed on any surface where the sunlight is optimized. There are two general types of solar cells. One is for charging batteries by supplying a voltage source. The other is used to generate heat, for example, to heat a pool or supply hot water to an electric generator. Solar cells are becoming more widely used as energy sources because they are easy to make and are low cost. Solar cells are most often connected in large arrays to develop a specific voltage and current level or to generate sufficient heat energy to power a turbine generator.

The most prominent solar battery was developed by Florida Power and Light near Tampa and was described in Chapter 5. Solar cell electrical power installations can be as small as a hundred square feet of panels on the roofs of trucks or homes where there is adequate sunlight. They can be as large as several thousand hectares known as solar energy *farms*. Solar cells used for direct electric energy applications connect the DC voltage to inverters that transform the DC voltage to AC and can be connected to the power grid or be used to charge EVS batteries. Solar cell energy is the least expensive of any form of renewable energy that does not pollute. The downside is they use huge land space which might be put to better use.

Astrophysicists have discovered only about 5% of the energy sources available in our solar system, and how the earth functions in our universe. There is a possible energy source based on the discovery of *neutrinos*. These are subatomic particles of three types or *flavors*, that possibly exceed the speed of light, thus defying the Standard Model of Physics. Scientists postulate that if the fourth flavor is found, it is a possible energy source in the future. Every aspect of life on earth is in constant change with new problems to be solved and discoveries made to help solve them.

6.2.5 Wind Energy

Another source of power is wind. Windmills have been used for pumping water, milling grain, and performing mechanical work for hundreds of years. Sails on ships move people and goods by wind power. Modern wind turbines convert kinetic wind energy directly into electrical energy that can be stored in batteries or connected to the electric grid. The first successful wind farm systems with over two thousand windmills were doing mechanical work in Denmark in the early 1900's. Electric generators were later coupled to farm windmills in the United States during the 1930s.

Modern wind turbine generators are gigantic complex machines typically anchored to 30 tons of steel reinforced concrete, or much more concrete when anchored to the seabed. Tower heights vary with the tallest over 500 feet high. Blade swing diameters are 750 feet on the largest wind machine such as the 13-megawatt General Electric Halide-X behemoths in Rotterdam, each of which can power a home for a full day in seven seconds. That is the same rate of energy as 12,340 homes for a full 24-hr day.

Figure 6.10 Renewable Sources of Energy

In the United States, the first large scale offshore wind farm began construction in 2017 about 15 miles east off the coast of Rhode Island with five 6-megawatt units three miles away from the thousand residents of Block Island. Initially, there were objections from a wide range of environmental concerns about whales and marine species to ruining the peace and tranquility of paradise. Over time, it was proven that marine life was enhanced, and that the remote historic island remained paradise, but now cleaner, quieter, and less expensive after the inland fossil-fuel power generators were removed.

Offshore from the upper Atlantic coast there was a different situation when 10 Humpback whales recently washed ashore on and near Lido Beach, New Jersey. Whales are carbon absorbers and contribute to combating climate change. Oceans sequester 30 to 50% of the CO_2 from burning fossil-fuel depending primarily on water temperature. Whales sequester 2 million tons of CO_2 per year, the equivalent of 225 million gallons of gasoline burned by ICE vehicles. An average single whale absorbs 30 tons of CO_2 per year compared to a tree that removes 45 pounds of CO_2 per year.

The use of wind is not free. Wind turbines are very expensive to design, manufacture, and erect. They need constant maintenance, create noise, have limited life, and in some places are an eyesore. The cost of deactivating and recycling a wind turbine is very high. In cases, the land could be put to better use because wind turbines need lots of land or ocean area.

Climate, or weather fluctuations, directly control wind flow so wind is not predictable at any given place over the long-term, especially on land. In the short-term, wind turbines are practical power options if the average wind is calculated to remain consistent for the design life of the turbine. Otherwise, the wind turbine may not be worth the energy and carbon cost to build and install it.

It is more sensible to erect wind turbines along certain coastal areas where winds are more predictable, sustainable, and generally stronger the higher the turbine tower. Careful thought must be given to the placement impact of wind turbines on maritime supply chain navigation, commercial fishing waters, and migratory marine life areas, but these are solvable issues in many regions. The most desirable locations for offshore wind turbines are as close as practical to the places of greatest power demand in major cities.

One of the most powerful energy sources are the polar jet streams that occur around the poles in the tropopause at about 11 miles above earth. These jets and ocean temperature are major factors affecting earth's weather phenomenon. The northern jet stream is responsible for tornadoes, hurricanes, and other violent weather anomalies in North America. Scientists have calculated that capturing only 1% of this energy is sufficient to power 100% of earth's total energy needs in 2050. However, science has only begun to contemplate how to harness this energy source.

While earth has abundant energy resources, it is becoming more technically difficult to use them as the population grows, as the individual use of electrical power climbs, and as air pollution control becomes more challenging. The situation is becoming more complex because the earth is populated by 7.8 billion people in more than 250 countries, each of whom have specific resources, climates, populations, wealth, cultures, and politics.

The estimated population will grow to 10 billion people by year 2050, This makes the need for astute energy management and pollution control more urgent. What occurs in one large country affects the pollution in other countries because air constantly moves around the globe in complex atmospheric circulations. In the United States, surface winds generally flow from west to east, but vary seasonally and widely in mountainous and desert regions.

With global temperatures erratically flip-flopping, winds will become unpredictable and more extreme. Hurricanes are becoming more frequent and violently destructive. Water supplies on land are decreasing as more drought is expected in southwestern areas of the United States, China, Africa, and northern India. Lightning storms continue to ignite the forests throughout the temperate zones on earth. There has been a recent decline of forest fires in western Canada until June 2023. and fewer trees with less underbrush.

As of September 2022, there have been record setting temperatures, unseasonably high winds, loss of harvests, and record numbers of forest fires in the southwest parts of the United States and in the European Union. Another summer of extreme forest fires is expected in the western and northern forests as the Artic warms. Parts of northern India have had extreme drought causing loss of grain crop yields and resulted in water rationing in Delhi, the world's second largest city. The population of India is expected to exceed China's in 2023.

Elsewhere in the southern latitudes there was record rain and flooding in Eastern Australia, while temperatures rose to levels never felt before as the Antarctician continent warms. Atlantic hurricane activity is moving northward in response to the polar jet stream unseasonably rising toward the North Pole. In the coming years there will be hurricanes in Boston and New Hampshire as the spiraling windstorms begin to shift northwestward after passing Bermuda. Predicting the future is nothing more than a guess.

Planet earth is undergoing a transformation of biblical proportions with tempestuous weather, inadequate power generation around the globe, and food and freshwater insecurity. Hostile earth conditions may continue to worsen until the world learns to prevent air pollution emissions and designs remedies.

Those who have never failed, never tried anything new.
— Albert Einstein

CHAPTER 7

ATMOSPHERE, OZONE, POLLUTION

E arth's biosphere and atmosphere consist of five expanding layers surrounding earth. These layers vary in altitude between the poles and the equator. This chapter is primarily interested in the first three zones which are: surface air, troposphere, and stratosphere. Of special interest is the ozone layer that exists above our surface air. These layers are vitally important to our existence and health. Pollution interferes with natural cooling, and the sun's ultra-violet rays will destroy life on earth without a filtration zone.

7.1 Climate, Weather, and Surface Air

Climate is a function of earths position relative to the sun, natural geologic activity, ocean currents, the primary northern and southern jet streams, and man-made air pollution. Normal climatic variation without excessive human air pollution is predictable in regions over about a decade based on location and seasons. Weather is the local effects of climate and pollution where one lives, what you see and feel outdoors.

From the beginning of the Industrial Age in 1870, the weather on earth has continued to worsen because of natural forces and human activity. In the year 2021, the amount of

CO_2 in the atmosphere was 4.21 parts per million, about 50% higher than it was at the beginning of the Industrial Revolution. The rise of CO_2 has not been linear over time, but exponential meaning that the rate of increase has trended up every year after 1870 with minor swings. There have always been natural periods of short-term extreme weather on earth. Global air circulation is complex and impossible to forecast beyond a few weeks to a month.

What is happening for everyone on earth is caused by human behavior and the unbridled use of crude oil. Planet earth since humans evolved has never been warmer than it is now.

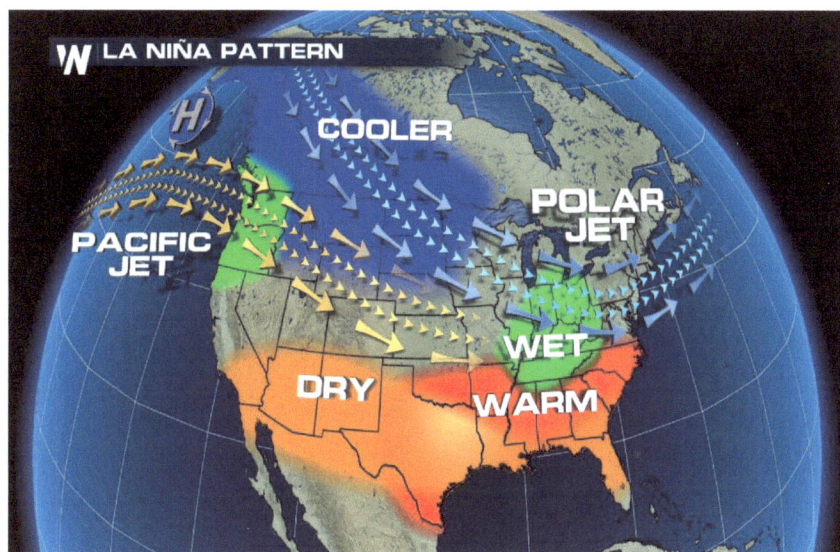

Figure 7.1 La Nina Circulation Pattern

Weather in the United States historically fluctuates between La Nina and El Nino circulations based on the Pacific Jet which is presumed to start in tropical Indonesia and is influenced by the northern Polar Jet Stream. Without the effects of pollution, the La Nina pattern is the result of cooler western Pacific Ocean water temperatures, typically as shown in Figure 7.1.

The opposite El Nino patterns are associated with warmer water in the tropical zones that causes a downward shift of the northern polar jet stream resulting in warmer temperatures in the north and wetter colder weather in the south. These two patterns tend to be consistent for a decade or more as they slowly cycle between the two pattern extremes. Air pollution can dramatically change that.

Air pollution has upset the circulation patterns with the polar Jet radically shifting downward off the west coast of California and circling back through northern Mexico to western Texas in an abrupt reversal of direction in early spring 2022. The overall effects of abnormal Pacific Ocean water warming and cooling plus pollution have caused both the Artic and Antarctic polar regions to warm faster than the mid-continent.

In the past few decades, the polar jet stream pattern has caused drought in most of California, the mid-west, and parts of Oregon and Nevada; record high temperatures in much of the upper United States; and an increase in forest fires in places where they are unusual, like New Mexico with a record land plant fire in 2022. In between, there has been monsoonal rain in Arizona, usually a drier state. Weather forecasting has become more complicated around the world. Similar conditions exist below the equator where seasons are reversed. This does not prove climate change, but it is an indicator. The past becomes clearer after the future arrives.

Earth's surface is warmed by 48% of available sunlight according to a Jan 14, 2009, paper written by Rebecca Lindsey of NASA Climate Science. The article reported that 25% of the heat energy received by earth is returned by evaporation; 17% returns from thermal radiation; and 5% is emitted back by air convection. Those numbers are changing, she implied because of air pollution.

Energy is never lost but exchanged in a continuous process that occurs in the surface air and upper atmosphere. Upsetting the natural balance is the cause of weather anomalies and apparent hostile climate change. The reason is that air pollution forms an unnatural blanket around earth causing weather warming or cooling in regions. This type of CO_2 insulation does not allow normal heat exchange, and it upsets the historic atmospheric circulations and warms both inland and oceanic waters reducing CO_2 absorption.

Figure 7.2 depicts the nitrogen cycle, the essential transition of the most abundant gas on the surface of earth and in the troposphere. Nitrogen is believed to be the sixth or seventh most abundant element in the Milky Way galaxy and is a major factor influencing weather. Air pollution and the ozone layer are discussed in greater detail later in this chapter. The ozone layer is essential for life on earth, without it nothing would stay alive.

The air between land and sea to about five miles, or 37,000 feet high is called surface air, the first atmospheric zone. Above this region and to about 11 miles high is where virtually all CO_2 pollution and greenhouse gases exit. The variable height ozone layer that shields earth is in the second zone.

7.2 Troposphere

The second layer is five to 11 miles above the surface air. The altitude and position of the tropospheric layer is not constant; it varies widely between the poles and the equator for complex reasons pertaining to the Pacific Jet and air circulation over mountains and across deserts. The air in this zone is 79% nitrogen, 21% oxygen, and 1% hydrogen with minor inert gases helium, neon, and argon.

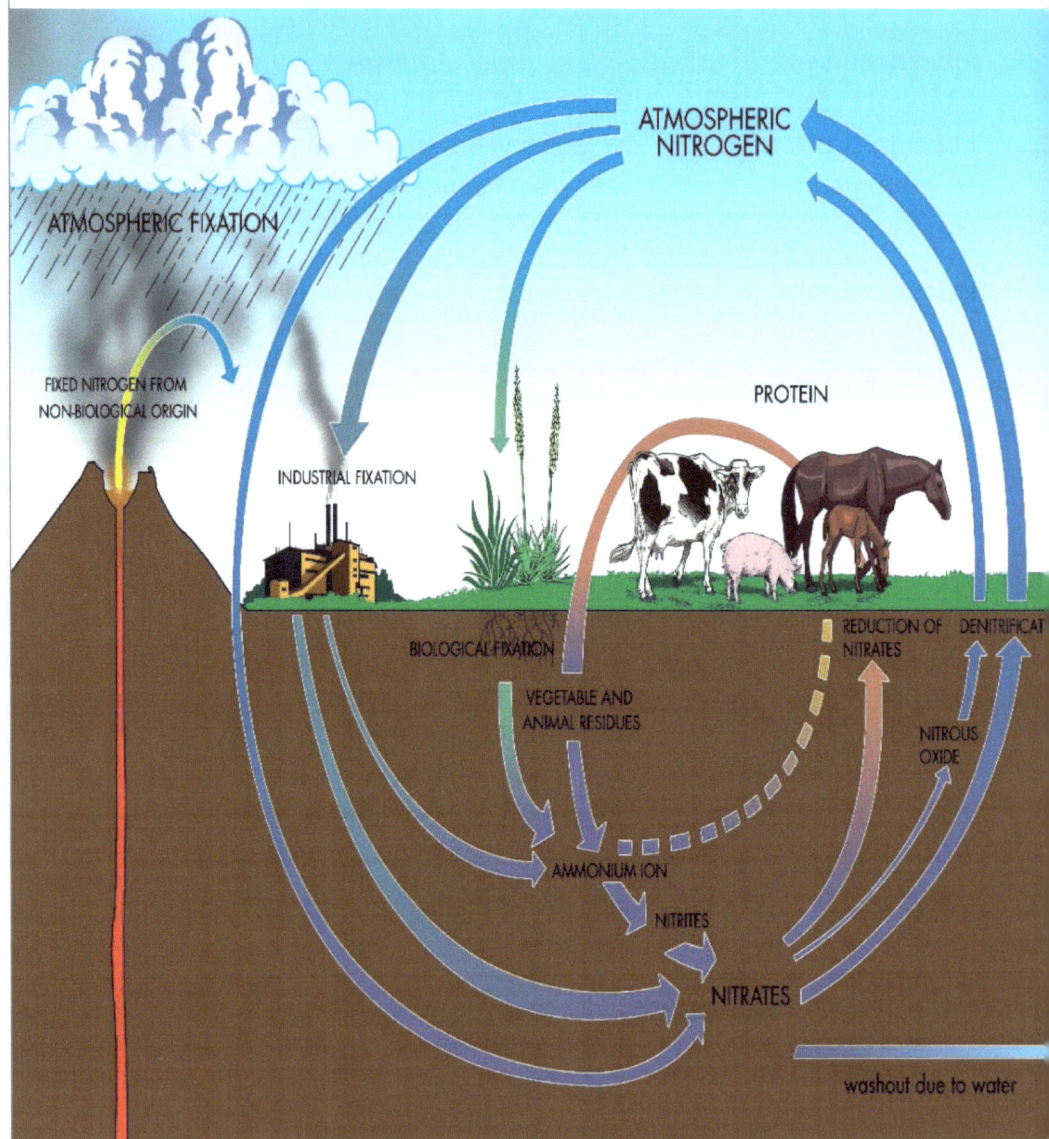

Figure 7.2 The Nitrogen Cycle Pictorial

Nitrogen binds with H2 to form ammonia, and with numerous oxides and chemical elements it combines to produce fertilizers that are critical minerals needed for plant life on earth. These include acids, nitrates, and sulfates, and N2 is a constituent in human and plant DNA. It is an essential ingredient in explosives, trinitrotoluene is TNT. It is also an inert gas which means that N2 is important in welding, metallurgy, and other industrial processes to prevent oxidation of metals.

The atmospheric flow in the troposphere is from west to east with interruptions and perturbations due to the Pacific and Polar Jet stream effects, several other kinds of flows, minor jets, and air pollution. These flows, and the jet streams, account for the major weather phenomenon on earth such as tempestuous weather warming, hurricanes, wind shear, sandstorms, blizzards, typhoons, tornadoes, and thunderstorms. The most violent and destructive to life and property are hurricanes and tornadoes.

7.2.1 Hurricanes

The most severe and destructive weather event is a hurricane, called typhoons south of the equator. The word hurricane was derived from the Caribbean language word, *Huracán* which means God-of-evil. They originate as tropical depressions, or low-pressure systems that move offshore from a large equatorial land mass like Africa, Australia and South American. Most of these depressions do not develop as they move into water that is not warm enough to increase their strength. Others generate high wind velocity because they move into warmer water that is the energy source of tropical storms. A hurricane develops when the cyclical wind velocity reaches 74 mph. The worst storm winds exceed 157 mph.

Storm winds are measured by instruments aboard an airplane that send data via satellites down to meteorologists who calculate wind speed using the Saffir-Simpson Wind Speed Scale program. Hurricanes are classified as Category 1 through 5, with Cat 1 when wind speed is greater than 74 mph to a Cat 5 with wind speeds greater than 157 mph.

The Atlantic hurricane season is from June 1 to November 31, and they are given human first names starting with letters A to Z. On June 1st, 2023, tropical depression Arlene was named on the first day of the season.

In the southern hemisphere, typhoon season in the western south Pacific is from January through June. Hurricanes have a low-pressure center called an *eye* that is free of clouds and where winds are calm. Hurricane tops break to an anvil shape like lower altitude thunderstorms; they vary from 40 to 55,000 feet high.

Hurricanes existed for centuries BC, but air pollution has apparently caused increased activity and greater strength since 1851. The five most deadly northern storms happened after 1990. In 2004, there were four hurricanes making landfall in Florida within a six-week period. The deadliest was Charley, a Cat 4 with a 7-mile-wide eye and wind speeds of 145 mph. Charley was exceeded by Katrina, a Cat 5 in 2005 that claimed 1,800 lives in Louisiana.

Many hurricanes followed until Ian, a Cat 4 In October 2022, with winds at 155 mph and a 14-mile eye diameter. Ian was just 2 mph short of a Cat 5, and it took at least 137 lives and was the 2nd costliest, most destructive hurricane in United States history. Ian was the worst hurricane in Fl since 1935. Another happened in 2022 and will happen in the following years until the planet is returned to natural historic atmospheric balance. That may take several decades because CO2 in the atmosphere takes decades to photolyze while more is emitted every second.

Most mid to long-range commercial planes operate between 35,000 and 39,000 feet, or six to seven miles high. Aircraft jet engines are powered by volatile kerosene jet fuel that emits CO2, water vapor, and heat; a combination that forms cirrus clouds called *contrails*. These tend to blanket the earth's surface holding down natural earth cooling and cause weather warming. Wind energy in the polar jet stream is enormous, with less than 1% of the available energy sufficient to power the entire earth in year 2050. Academics have studied the jet stream, but the technological development to harness this energy is decades away, if ever. Polar jet streams are high above the surface air at roughly 11 miles high in the United States with wind speeds of over 275 mph.

On earth, the temperature decreases with altitude to the top of the troposphere. Above this layer is a zone known as the *tropopause,* wherein a temperature inversion occurs, and atmospheric *fixation* starts to end. With increasing altitude above the tropopause, the air becomes warmer, not colder as we experience when rising from the ground on earth. There is also ozone in the troposphere, but it is minor compared with the stratosphere.

7.2.2 Tornadoes

Tornadoes are spiral or wedge-shaped storms, the second most violent and destructive climatic force on an individual basis. There are 60 times as many tornadoes as hurricanes in the United States, and they exceed the annual cost of hurricanes because there are so many.

In the United States they frequently start along western Texas and Oklahoma flowing eastward in a region known as Tornado Alley, but they can begin to form anywhere. Unlike summer hurricanes, a tornado can happen in every season at any time in most countries with furious winds, rain, snow, hail, and lightning. Tornados are classified on the Enhanced Fujita scale from F0, winds over 73 mph to F5, winds 261 to 318 mph. F5 tornadoes often exceed the wind velocity of a Cat 5 hurricane lifting cars and objects and destroying homes and medium sized buildings while killing dozens of humans a year.

The most violent and destructive hurricane in the United States was an F5 that took the lives of 695 people in MO, IL, and IN in March 1925. The most recent deadly tornado happened in April 2023. It took 27 lives in Mississippi. The last two years during an active climate change period, the number of hurricanes in the United States has gone up while the number and severity of tornadoes has slightly decreased. Climatologist report that the reason is inconclusive, but climate change will impact tornados in an unpredictable way.

7.3 Stratosphere

Above the tropopause is a layer from about ten miles to thirty miles high that varies between the poles and the equator. In this region is the ozone layer and Green House Gases CO_2, CH_4, N_2O, O_3, and water vapor absorbing 98% of the sun's ultraviolet radiation protecting life on earth. The CO_2 released from earth accounts for three-fourths of total GHG's.

At the 2020 rate of CO_2 emissions, the average earth temperature is expected to rise by 2.6°F in the year 2050. Meteorologists are now saying that the estimate is too low. While over long periods of time, earth's temperature has widely fluctuated from Ice Ages to tropical warmth periods, earth has not been this warm in over three million years. Global warming is likely a function of man-made pollution and the consequential overuse of earth's fossil-fuel resources.

The *Greenhouse Effect* is the thermal energy exchange between the sun and earth's radiation that results in an average earth temperature of 59°F. Without this effect, the average temperature on earth would plunge to 0°F. However, because of air pollution, the world-wide temperature is rapidly rising due to reduced surface cooling because CO_2 insulates earth.

7.4 Ozone Layer

The purpose of the ozone shield is to protect earth from the sun's ultraviolent radiation. Without this zone, UV would destroy plant and animal life on earth. High intensity, medium frequency UV results in cancer, loss of sight, depleted plant tissue, and other health problems. Ozone is dispersed in the stratosphere between nine and 22 miles high, peaking at about 18 miles at poles and equator, so it does not protect every country equally. Lower latitude countries are normally warmer than the higher. Most Latin-America countries and are warming more than usual.

To explain how the ozone process works, think of the sun above earth and in between is an active closed loop process. The first step is water vapor and O_2 molecules are released into the stratosphere. The second step is O_2 being photolyzed or split in two by the sun's UV radiation. The free oxygen atoms quickly bind with nonphotolyzed O_2. This process converts diatomic O_2 into triatomic O_3 called ozone. This happens slightly above most other polluting gases.

Methane, CH_4, a reactive greenhouse gas is also photolyzed by the sun at a decay rate up to nine times faster than CO_2. The process continuously repeats in the closed loop atmospheric system s shown in Figure 7.3

Ozone is an unstable molecule on earth, but in the stratosphere it is long-lived. UV splits O_3, but it recovers. However, O_3 is destroyed by these molecules released from

Figure 7.3. The Ozone Process, CO2, H2O and CH4

earth: chlorofluorocarbons, bromides, sulfides, chlorides, and nitrous oxides. CFCs were widely used in air conditioners and aerosols until about thirty-five years ago. Now these gases have been banned in most countries. Ozone depletion is a partial factor accounting for the weather extremes of the last decade, but it is not the only reason.

Carbon dioxide is the major cause of air pollution and weather warming, but it is a minor factor in depleting the ozone layer. CO2 forms an inconsistent, broken boundary layer over earth that prevents warm air from escaping the surface and holds down other pollutants causing smog. It is the primary cause of earth's tempestuous weather changes which causes unusual warming and cooling, and other anomalies, including more wind shear, hurricanes, and tornadoes. In short, it upsets the normal *balance* or *carbon neutrality* of the atmosphere.

The most abundant gas molecules on earth, N2 and O2, are not affected by the greenhouse effect, but various pollutants cause serious human lung and respiratory problems. Tumultuous weather affects everything from the world food supply, transportation, production, and inflation, to loss of animal and human life and emotional disorder. Most worrisome, it is caused by past and present human behavior and corporate cupidity regarding a declining valuable energy source, fossil-fuels.

7.5 Pollution

Pollution is simply undesired, wasted, or unusable energy including garbage, scrap metal, discarded plastics, and both natural and made-made emitted gases. There is an enormous amount of pollution on land, in the oceans, and in lakes and waterways. During the last 250 years, humankind has carelessly used even the most secluded parts of earth as a garbage dump. With 7.8 billion people on earth in 2021, garbage can be found from the highest point on earth, Mt. Everest, to the deepest point of the ocean, the southern part of the Mariana Island Trench near Guam.

Handling of waste, purification of water, revitalizing the oceans, and cleansing of the air are major issues and expensive problems to solve. They are the result of human carelessness and over-use and mismanagement of fossil-fuels that were created 300 million years ago when a gigantic asteroid collided on earth forming the Gulf of Mexico. In this book, the primary concern is internal combustion exhaust, smoke, fumes, smog, greenhouse gases, and pollutants that are released into the surface air and low troposphere from burning coal and crude oil.

The cost of all pollution is unfairly distributed throughout humanity both locally and globally because of negative *externality*. This occurs when the pollution produced by manufacturing or production of one company or country affects the well-being of another business or country that is not compensated by the polluting entity for damages to their livelihood. Another example of externality exists when a person buys a high-cost, high horsepower, 1000+HP personal vehicles using increasingly scarce energy resources while expelling wasted energy in the form of air pollution at the expense of those buying low-polluting conservative cars. These cars should be pollution taxed.

Luxury EVS manufacturers dispute this viewpoint claiming an EVS uses the same energy as low horsepower cars when in traffic moving at the same speed. They discount the fact that luxury EVS pollute the air from the start by mining more lithium than needed which causes air pollution before the car is even made. Luxury cars weigh more, use high-cost material. and use more energy to recharge the batteries.

MAJOR CAUSES OF AIR POLLUTION

Fossil-Fuel Electric Plants and Refineries	32%
Transportation of Goods and People on Land	26%
Agriculture and Livestock	11%
Ocean Freight Shipping	10%
Manufacturing and Mining	10%
Air Transportation	6%
Computers and Other Causes	5%

The percentages listed above are estimates for the United States in 2021 and vary widely around the globe by country, geographic location, and by whom the estimates are made. It is important to be aware that air is dynamic and continuously circles around the earth in massive polar and oceanic jet streams and in less forceful circulations. Everything is dynamic on earth and changes constantly, both in the atmosphere and on the biosphere.

The health of every person on earth is affected by pollution. Therefore, every person must be actively involved in, and acquire an understanding of earth, energy, power, pollution, and the cost of hypocritical actions. With the world population estimated at 10 billion by the year 2050, everyone must become educated in, and become energy and pollution aware. According to the Universal Thermal Climate Index, air temperature of 38°C is considered very strong heat stress, and at 46°C it is extreme or near death for at-risk people. A Centigrade to Fahrenheit conversion table is shown in Figure 10.1 in Chapter 10.

Examples of global earth events that affect everyone are the war on Ukraine by Russia, and record 113°F extreme heat in northern India. Both events caused worldwide grain shortage because of reduced yield, fire, and air pollution made by explosives and unusual dusty winds. In April 2022, prime minister Narendra Modi of India quipped that world climate problems cannot be solved without India, the second largest country in the world with 1.4 billion people. Within the next two years, the population of India will exceed that of China, both becoming crowded countries that are responsible for a high percentage of world air pollution.

The 2021 global conference for climate change in Scotland was attended by *climate experts* who arrived in 400 private jet aircraft rather than taking commercial flights. This was not a good example of clean air awareness and practices by climate experts. Personal jet air travel accounts for 1.5% of global air pollution now and is expected to become worse as the wealthy population grows and continues to buy personal jet airplanes.

At the World Economic Forum held in Davos, Switzerland in May 2022, over 1,000 influential economists and climate experts arrived in hundreds of polluting private jet planes. While there, personal electronic *tags* were proposed to be attached in all humans to monitor their daily eating and movement activities. The purpose was to monitor habits of the average person and to collect data for climate change control of the masses. Beware of what you believe from climate change activists who predict a climate crisis.

The premeditated war of Russia on Ukraine brought to attention another problem in the Artic, a region that Russia intends to dominate. Both the Antarctic and Arctic are beginning rapid warming. Ice layers in Siberia and regions around the north pole are covered with ice, called *permafrost* up to 500 feet thick, and under them are methane and natural gases. Surface warming and melting causes water to seep into cervices which enlarge to wide, deep cracks in the ice. This releases methane into the surface air that causes more warming in a cyclical closed-loop process.

Methane gas can be detected from satellites and hopefully be captured and converted to climate-moderating hydrogen fuel in the future. The consequences for not doing so could be destructive to animal life and humankind. Neither humanity nor rabbits and sharks can flourish with only ideas in an unhealthy environment. Something must change that.

7.6 Future Weather

It is not possible to forecast future weather beyond about a week or two. Predicting the future climate is even more difficult. Meteorologists can assume seasonal changes based on historical averages, but nothing about earth is certain. Everything changes, the universe and earth remain in a constant state of flux since the beginning.

Historically, California has had mega floods following a decade or so after by mega droughts. Because of population expansion and other factors, if history repeats, there could be disastrous rainfall for that state. It is possible that parts of the state will become more islands. Another major concern is rapid warming of the poles, both of which were once tropical. Polar warming affects worldwide climate and local weather anomalies, including drought, forest fires, and hurricanes.

Drought has caused most rivers in the United States to flow less water. The Mississippi and Colorado Rivers are at extreme low levels which contribute to present and future food scarcity and inflated prices because of reduced irrigation. Ships and barges on the Mississippi River are confronted with the lowest water depths in 20 years restricting low-cost shipping of food products. The price of agricultural and transportation diesel fuel rose due to closure of crude oil pumping on federal land worsening the looming food insecurity in the United States.

Hostile weather is now rapidly changing all parts of the world. In Sichuan Province, China, one of the country's largest manufacturing regions, drought in other parts of China have reduced water flow of the Yangtze River to record lows curtailing electrical power generation by 50%. Major rivers in most parts of the world are flowing less water because of drought. Forest fires in every continent except Antarctica have caused enormous air pollution and worsened drought because of evaporation and loss of millions of acres of carbon capturing trees and plants.

War in Ukraine and drought in India and parts of Africa and South America contribute to the global instability of weather and a possible future climate change crisis. The opposite is happening in other places on earth. Floods in Nigeria and Pakistan have killed thousands of people. Only through a global effort to moderate air pollution will the world climate stabilize.

*Water and air, the two essential fluids
on which all life depends, have become
global garbage cans.*
— Jacque Yves Cousteau

CHAPTER **8**

MAJOR CAUSES OF POLLUTION

I n Chapter 7, the sources of pollution were mentioned. This chapter examines these in greater detail with emphasis on power plants, freight trucks, cargo ships, automobiles, and aircraft. Fossil- fueled crude oil, coal, and natural gas powering electric generating plants are the major pollution sources on earth at 32%, with transportation of people and goods causing about 27% more of earth's air pollution. These two are the major carbon dioxide and noxious air polluters.

At the beginning of the year 2022, the population of earth was 7.8 billion people. Everyone breathes oxygen and emits CO_2, requires food to eat, uses transportation in some manner, and generates pollution as waste in various forms. A list of all pollutants on earth would be quite long, and most people know what they are. The less significant and natural pollutants that we cannot prevent or control, for example lightning caused forest fires, the oceans, and volcanoes are omitted in this chapter. They account for only about 2% of pollution and tend to be uncontrollable. Some forest fires can be prevented by better forestry management and clearing of underbrush, but that topic has been debatable among foresters for over 100 years. The debates were probably based on economics, not on climate change or air pollution.

What is meaningful is that air pollution is the primary cause of death for over five million human beings per year. Two million of these occur in India, with most within 50 miles of Delhi. Air pollution is one of the major causes of respiratory diseases. There are scores of people in India who have never seen the sun or moon, much less stars, because of smog and foul air. This is also the case in parts of China. Both countries are the greatest air polluters on earth because they burn the lowest grade of lignite coal without carbon capture techniques. They are not the greatest polluters per person.

Atmospheric air moves constantly. Bad air in one part of the world circulates to other parts of earth. Air pollution in Peru, SA was chemically traced to the cause in Germany. Controlling pollutants and reducing the use of fossil-fuels is a global problem that requires the cooperation of every country and person on earth. Local pollution measures are helpful, but without global participation the apparent climate will continue to deteriorate until the major causes are addressed and resolved.

8.1 Carbon Cycle

Figure 8.1 depicts the carbon cycle, and as illustrated, hopefully clarifies the overall theme of this book. The word *cycle* means that carbon is exchanged rather than eliminated in the natural processes on earth and in the atmosphere. Carbon is a vital element.

The word *carbo* is Latin for coal, and in solid form includes lowest grade lignite, bituminous, and various grades of hard coal, graphite, and diamonds. Anthracite coal is a key ingredient in steel making and manufacturing. In gaseous form carbon is a component of CO, CO_2 and CH_4. Carbon seems to be part of many things, but it makes up a small fraction of 1% of the chemical elements on earth despite being part of all fossil-fuels.

In the universe it is the 4th most abundant chemical element. The noxious effects of gaseous carbon on earth belie its quantity.

Everything on earth, including all living forms, and earth and the sun eventually decays and dies. Animal waste is in the form of fluids, solid excrement, methane, and minor CO_2. The animal excrement fertilizes the soil, and in the longer-term converts to fossil-fuels. The belched gases from farm ruminants are mostly CH_4 that converts to H_2 in under ten years. Excess CO_2 forms a moving patchy *blanket* around earth that causes global warming, and upsets earth's natural cooling system and the historic geological and atmospheric balance.

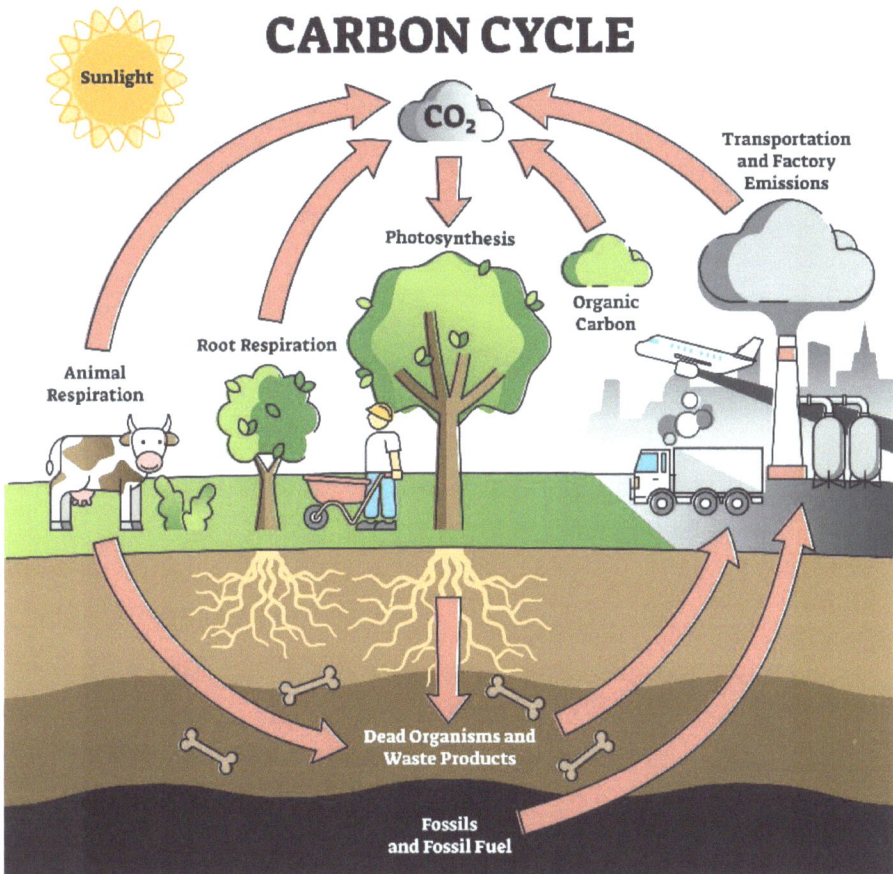

CARBON CYCLE

Figure 8.1 The Natural Carbon Exchange Cycle

The next process in the cycle is trees and plant life that need CO_2 for photosynthesis in which photons convert the CO_2 to O_2 and glucose sugar that feeds the bats, birds, and bees as they pollinate plant life. Their roots require chemo-fertilizer and CO_2 that can be routed from polluting sources to nourish trees and plants. Without CO_2, all plant life on earth would perish.

Fossil-fuels in earth power cars, trucks, and aircraft that are organic carbon emitting sources including CO_2, dust, fine rubber particles, and freon from air conditioning. The same types of fuels power factories, buildings, huge machinery in mines, and electric power generating plants.

The cycle would not work without the sun, the ultimate source of power on earth. The cycle continues efficiently if pollution is balanced with O_2 production and natural earth surface cooling. As the air pollution increases, the earth warms, weather turns violent in places, and the health of all living matter on earth declines.

8.2 Power Plants

Fossil-fueled power plants are the *major sources of pollution on earth*. Humankind has used coal for small fires since centuries ago and coal is burned today in gigantic power plants around the world. It is abundant on earth, easy to mine, and is an economical source of electrical power generation. But when coal burns, it releases free carbon which instantly binds with O_2 to form CO_2, the worst polluter in the world.

According to Fuel Cell Energy Incorporated, the manufacturer of fuel cells and carbon capture systems, fuelcellenergy.com, a 500-megawatt coal-fired electrical power generating plant emits 3.6 million tons of CO_2 per year. A ton of CO_2 means about 3.7 times the weight of the coal burnt because O_2 binds with C in combustion. This amount is equivalent to the CO_2 emitted by 685,000 average gasoline powered cars in a year.

$$Coal \longrightarrow CO_2$$

Atomic weights: $\quad C = 12 \quad\quad O = 16$

$$CO_2 = 12 + 16 + 16 = 44$$

$$CO_2/C = 44/12$$

$$= 3.67 \sim 3.7$$

The need to replace coal by designing and building alternative energy power plants is obvious. Pressure on China, India, Russia, and several other countries to comply may not be successful in the near term because of wide economic, cultural, ideological, or political differences.

Figure 8.2 Power Plant and Refinery Pollution

Power generation plants burning coal and crude oil products account for 32% of the pollution on earth, and when coupled with cargo ships, land freight, air travel, and automobile transportation, make up about 95% of the pollution on earth. There are alternative fuels available, as well as means to reduce CO_2 emissions, but they may not be economically feasible in many countries. Immediate steps to moderate pollution are necessary, but crude oil is a vital material for other essential products. The use of crude oil needs to be moderated, not ceased, and its gas byproducts better engineered and managed.

There is no need to take drastic action to curtail crude oil pumping. The consequences of doing so include inflation and unemployment of highly educated, trained engineers and science professionals. Crude oil and gasoline prices have risen rapidly in 2022 because the supply of crude has decreased while the demand increased. This results in price increases for everything that is not produced at home and needs to be transported. Because crude oil is the raw material for so many other products and purposes other than electric power generation, it is a vital commodity. The transfer to alternative power sources must be carefully considered, planned, managed, and executed by industry and governments with serious thought as to the unintended consequences of making major bad decisions.

The amount of crude oil reserves on earth is huge, but not endless. At the beginning of the year 2020, the world consumed 35 billion barrels of crude per year. The total oil reserves on earth that are financially recoverable were estimated at 1.7 trillion barrels in 2020, with the leading oil reserves by nation in billions of barrels were Venezuela 304, Saudi Arabia 267, Canada 170, and Iran 157.

The United States was Number 11 at 48 billion barrels, enough for about seven years at the 2020 consumption rate not counting the Strategic Petroleum Reserve and some deposits beneath federal land and offshore. In 2021-22, about ½ the SPR was removed from the national security supply and some of it was secretly sold to China, an ideologic enemy. The total world supply is estimated to last for 45 to 50 years, assuming the 2020 rate of consumption. The use of crude oil around the world must continue to decrease, not just because of pollution, but at the current rate of consumption it will not last much longer for future generations.

8.2.2 Petrochemicals and Plastics

Crude oil is boiled at up to 600°C and distilled in large columns and cooled to produce various grades of petrochemicals as fuel and to produce numerous grades of feedstocks to make plastics. How that is done is like making moonshine. In general, all the common power fuels come from cracking crude oil hydrocarbons formed in the distillation columns and extracting it at different temperatures.

The six fundamental plastic stocks are processed further from ethane and butane to yield the six key plastics which are: ethylene, propylene, benzene, toluene, xylene, and butylene. Plastics are made from monomers that can bind together with identical monomers and bundled to form polymers. There are natural polymers like proteins and glucose, and synthetics like ethylene (Teflon), vinyl chloride (PVS), and styrene. Making plastics sends CO_2 and methane into the atmosphere and creates land pollution.

The problem is not only making plastics but how to recycle or dispose of them. Plastic food containers enabled the *throw-away culture* that prioritizes convenience over pollution resulting in 380 million tons per year of single use plastics waste. That is more than the weight of the entire human population. Single use plastics do not breakdown after use, they break up into microplastics that are not recyclable, ending up everywhere.

The most common plastics are polyethylene tetraphene (PET) for food and beverages and polyvinylchloride (PVC) for plumbing and credit cards. PET is recyclable, and PVC and polystyrene (PS) used for egg cartons and take-out foods are not. The most dangerous plastic to humans is Styrofoam, an expanded type of PS.

The plastics industry is gigantic. Practices, procedures, and restrictions are under continuous worldwide review. In the year 2023 there is a Global Plastics Treaty underway intended to set policy and restrictions. But telling other countries what they can and cannot do about plastics is only effective if imports of dangerous goods from non-members are banned and enforced.

8.3 Mining

The search for and removal of minerals from earth uses energy from power generating plants and fossil fuels. Mining operations are dusty, disturb the surface, and create land and air pollution. The most used and processed materials are sedimentary minerals limestone, sandstone, and shale. Rocks of most kinds to chalk and mud are mainly calcium carbonate left during the formation of earth when they settled in water. The most widely used materials around the world are limestone rocks mined to make cement and concrete. In doing so, they create over 4% of air global pollution. Cement is the essential ingredient in concrete that binds sand and aggregates together for high compression strength and when reinforced with steel are the foundation and structural components in large buildings.

Cement is made in plants where raw limestone is crushed from quarry-size limestone to small 3-inch size and fed into giant rotating drums where it is tumbled with ash and heated by natural gas to 2,700°F. The output is marble-size balls called *clinkers*. After cooling, clinker is crushed to fine powder and packaged in sacks with a trade name and usually called Portland Cement. Part of the cement is converted to concrete mixing at the plant and the rest are shipped by trucks, trains, or ships to end users and distributors. Every reasonable effort is used to minimize dust, but the operations use both energy from the electric grid and great amounts of natural gas. The final product is the most vital solid resource on earth.

Figure 8.3 Cement Manufacturing and Concrete Mixing

Shale is a softer form of calcium carbonate found on the surface and deep underground that traps ancient life and plants that became deposits of crude oil whose extraction is a form of geology and mining. Crude oil is mined in a manner familiar to most people. It is pumped from underground deposits to large tanks for short term storage and transported to refineries by truck, rail, ship, and pipelines. Energy and pollution occur in making the devices used to find, tap, pump, and deliver crude oil and natural gases to users. Mining is an essential matter that does harm to the world and better resource management and usage is needed.

8.4 Manufacturing

The originators of climate change were those who developed and produced new and improved goods and processes essential for a growing population. Climate was not a problem 250 years ago, but as the population grew increased manufacturing of goods and the development of new materials and techniques became essential. Eventually, heavy manufacturing of concrete, timber and steel for buildings and highways became urgent. It was only a matter of time before the atmosphere was filled with CO_2 and other pollutants that created today's climate change. Chapter 9 discusses means used to moderate pollution in these industries.

Producing goods for personal use and household products is minor compared to refining oil, making steel, building huge structures, and many small products are imported. The exporters are responsible for the pollution of those products. Chapter 10 is devoted to moderating the air pollution caused by massive transportation and importing and exporting manufactured products and raw materials.

8.5 Buildings and Homes

Skyscrapers and towers use huge amounts of power and natural resources to build. They often are built where another building has been torn down creating waste and air pollution. While many are architecturally beautiful, they are inefficient in their use of space, lighting, heating, and cooling. In a June 16, 2022, article about the topic, *The Economist* wrote, *the carbon footprint of buildings is growing and expected to double by the year 2050.* The article went on to state that the construction industry is horribly climate-unfriendly accounting for about 40% of worldwide air pollution emissions.

The energy needed in heating, cooling, and lighting these large energy consumers cannot be supplied alone by renewable energy sources. This is a major problem because by 2070, the world will be out of oil, with the assumption that the 2020 rate of consumption will continue. In cities where most of the gigantic buildings are found, there is no room for planting trees or landscaping for CO2 capture and oxygen generation. The largest cities in the world are already covered with vehicle exhaust and air particles. A notable exception to the trend is in Singapore, a small Asian city/country known worldwide for futuristic planning and environmental engineering.

The homes of those who live closest to the largest cities face a problem from breathing the foul air. In addition, they become the source of more energy demand because of personal computer usage and lifestyle while not contributing anything to help clean the planet. As a minimum, owners must be encouraged to plant trees and shrubs to supply oxygen. Where adequate sunlight is available, solar panels should be used to recharge EVS and computers to reduce drain on the electric power grid.

Figure 8.3 Trees and Plants on Singapore Buildings

8.6 Heavy Land and Air Freight

The movement of materials on land from where goods are produced to where they are used is a major source of pollution. Trucks and trains use crude oil and coal for locomotion and are significant polluters. Electric trains are no different since in most cases the electrical power source is coal or crude oil power plants, except in a few countries with ample hydro-electrical power. Trucks also emit rubber tire particulates the size of human hair diameter into the air. Most long-haul trucks have eighteen large tires moving at the maximum speed limit. This emits fine particles which create smog, foul air, and is dangerous to breathe for humans and animals. It is most abundant and noxious near large population centers and major airports.

Freight transportation vehicles have an average yearly mileage range of over 65,000 miles per year, five times that of automobiles. They burn diesel fuel which is much more polluting than gasoline for personal automobiles. The alternatives for long range transportation energy are batteries or hydrogen fuel- cells. There is also advanced technology that collects particulate matter from tires. These are available now, but they are expensive and not legally required for 18 wheelers.

Heavy aircraft importing and exporting goods burn kerosene, a highly refined hydrocarbon. When ignited, the jet fuel releases not only CO_2 and water vapor, but fine particulate matter which causes smog near cities and can affect health. Maximum power is generated from planes at take-off and landings close to large cities where human populations are most susceptible to lung diseases. At cruising altitudes, they emit water vapor known as contrails which form cirrus clouds filled with CO_2 that hinders normal surface cooling.

8.7 Automobiles and Aircraft

Cars, small delivery trucks and aircraft not only emit CO_2, but their tires continuously release tons of fine rubber particles into the surface air. These fine particles are invisible individually, but in mass create smog that is dangerous to breathe. Rubber tires are manufactured from crude oil whose wheels emit both CO_2 and methane. This is a critical situation because without crude oil the only other source of rubber is from rubber trees in Asia which could not supply even 0.1% of the rubber tire demand on earth.

Commercial and private jet planes emit CO_2, nitrogen oxide, and water vapor. The mixing of CO_2 and water vapor form cirrus clouds at the principal flight altitudes between 35,000 to 39,000 feet. The Boeing B-737 short to middle range plane holds over 40,000 pounds of kerosene jet fuel that is burned in less than eight hours, often over the same flight routes on the same day. Cirrus clouds produced by jets are loaded with CO_2 that stays in the atmosphere for decades. Heavy planes pollute even more than B-737's. Air pollution is especially bad in large city residential areas within about 10 miles from a major airport.

The use of private jet planes has dramatically grown in the last 75 years as the number of wealthy people on earth has increased. These aircraft pollute at the expense of those

Figure 8.4 Heavy Freight Hauling Causes Air Pollution

who do not travel, or who book commercial flights, and in a fair world they should be assessed a carbon tax. Private aircraft emit many times as much air pollution per person as commercial aircraft.

There are over six times more cars, pickup trucks, and delivery vans than freight trucks worldwide. Wikipedia articles estimate there were about 284 million vehicles registered in the United States alone and 1.45 billion worldwide with about 363 million total commercial vehicles. Some countries and states have begun issuing dates as soon as 2033 when new fossil-fuel car registrations are not allowed. With only three million EVS in the United States, it will be difficult for car manufacturers to produce enough new cars to meet the demand. It may be more climate friendly to stop licensing older high horsepower cars and diesel pickup trucks that tend to be more polluting than new ICE vehicles.

Horsepower of personal automobiles and planes has steadily increased since the first ones were invented. In the case of cars, there is no practical reason for excessively high horsepower except for illegal racing, police, and emergency use, and showing off wealth or status. High horsepower cars pollute the air. It makes no difference whether the auto is an EVS or an ICE car, the higher the horsepower the more the pollution from getting the raw materials to refueling or recharging. A carbon tax should be placed on vehicle power in a fair-minded, clean-air world, if the cars are driven. There is nothing wrong with owning antiques and older high-power vehicles if they are rarely driven.

In the leading high performance car producing countries, most notably the Germany, England, Italy, and the United States, personal car and pickup truck high horsepower is more cultural than practical. There is no logical reason for high car horsepower except for commercial pickup trucks and some large vans. All energy to power conversion exchange is less than 100% efficient and contributes to air pollution. The pollution solution is for mass voluntary moderation, but not socialism, around the world. Even the wealthy and upper class can learn how to moderate pollution, but humans are often slow to alter attitudes, behavior, and cultural habits.

The IEA and many organizations favor carbon taxes on high horsepower personal vehicles, boats, and planes. Industrial and commercial polluters could be warned to moderate over a reasonable time or be carbon taxed. The principle of externality infers that the wealthy and those who are energy extravagant are not entitled to pollute at the expense of everyone else living on earth.

Climate change activists who initially supported the unprepared, urgent rush to manufacturer EVS with government tax incentives should have tempered their agenda to a more conservative management of fossil-fuels. As of June 2021, the administration in the United States has maintained its position of limiting fossil- fuel access and put accelerated pressure on car companies to manufacture expensive EVS that only the upper class can afford. This resulted in the price of new and used gas-fueled autos to shoot up by 22% for the middle and lower class. In just over one year, the price of gasoline has doubled hurting retired folks, middle-class commuters, and the lower-class the most. Inflation quickly reached the highest level in mid-year 2022 since 1980 and may remain elevated for years.

The result prompted General Motors to slash the price of its $32,500 2023 Bolt EVS by $6,000 from the 2022 sticker price. The federal government enacted a $7,500 reportable income deduction as a tax incentive to buy an EVS. The problem is that did not help the lower to middle class families and retired who pay no or very little annual income tax. They are confronted with surging gas prices and expensive used gasoline car prices and need an affordable $26,500 EVS car.

Chevrolet also announced its Equinox EVS in September 2022 touting it as the EVS anyone can afford. Perhaps so if anyone can afford the cost of recharging it. Ford admits that EVS will cost more than their ICE counterparts for many more years and that a low-cost EVS cannot be produced by the major United States vehicle manufacturers for a profit.

Meanwhile in China, similar low-price EVS have been mass produced for more than 16 years. Since 2008, Tesla has manufactured three million EVS for sale in China, Europe, and the United States, but not many are low-priced even though they are simpler to build.

8.8 Computers and Artificial Intelligence

Fifty years ago, in 1972, an Intel 8008 microprocessor, shown in Figure 7.5, cycled at 20kHz with an 8-bit data bus and an addressable memory of 16 k bytes.

Figure 9.5 Intel 8008 Microprocessor (1.1 inch wide)
Photo: Konstantin Lanzet–CPU Collection Konstantin

Today there are more than a billion computers from inexpensive cell phones to ingenious smart phones and home desk computers to gigantic super computers that fill a large room and require considerable electrical energy to power and cool them. Computer bit rate has advanced from milli-bits to tera-bits per second with photo etching nano technology and wonderous sub-microscopic chemical line printing and connecting processes.

Microprocessors have followed Moore's Law, a prediction by Intel co-founder Gordon Moore that every two years microprocessors would become cheaper and faster with twice as many transistors. His empirical theory has proven correct until 2022, when removing the heat from more than a million transistors in a chip used more energy than the equivalent computing energy. Today's challenge is to create transistors that approach the size of atoms with tiny printed electric lines that cannot be seen by the human eye using a common laboratory microscope.

In July 2022, Micron Technology announced a NAND chip (an output signal when inputs to a logic gate are not all the same) with 232 layers of memory cells totaling two terabytes of data processing in a package smaller than one-half of a postage stamp. The amount of data processing capability is more than 100,000 times greater than an Intel 8008, in a package about one-quarter the size. That comparison confirms Moore's Law even though one is a microprocessor and the other a large memory chip.

The beauty of microprocessors is that they can be applied to almost any task unlike single use inventions like home fixtures and small appliances. The downside is that with growing computer usage and planned data center expansion, global electric energy production will have to ramp up by about 5 to 7% by the year 2030. There will continue to be an ever-increasing demand for computers and renewable cleaner electrical energy generation to power them in part because of artificial intelligence.

AI is an emerging application of computer machines capable of processing by inferring, perceiving, and synthesizing information at an extremely high rate. AI machines use massive amounts of acquired data from humans, commerce, and nature to think and control functions like a human or groups of humans. Major world changing inventions and technological advances will occur. AI is capable of miraculous goodness, curing diseases and 3D printing of human organs. In theory, AI can possibly cause unimaginable human destruction or nuclear annihilation.

Computer engineers and information technology scientists believe it must be regulated because of potential worldwide harm.

Doing that on an international basis will be challenging, if not impossible. There will be debates on how to do it. China intends to become the world leader in AI technology. With industrial spying and global access to technical information, it may not be possible to stop a rogue country from using offensive AI.

As far as climate change is concerned, AI may hasten the development of nuclear fusion power plants for electric energy without harmful emissions. It is also possible that one country may interfere with another country's air pollution solutions because they apply to national defense, global economics, and the vital resources, pharmaceutical, and semiconductor industries. The question is, what is the essence of AI?

8.9 Wars

The late February 2022 premeditated invasion of Ukraine by Russia represented not only a gigantic waste of energy, but the release of massive air pollution in the forms of CO_2, dust, glycerides, nitrates, nitrites, sulfides, and currently unknown other pollutants. The ruthless targeted destruction of civilian facilities, including schools, housing, and even hospitals has been unprecedented in the history of humankind. Ukraine was never a threat against Russia. The civilian atrocities may have been the Kremlin's miscalculated attempt to disguise the true weakness of Russia and cowardice of Vladimir Putin.

Massive tons of trinitrotoluene, nitroglycerine, and incendiary explosives in the forms of missiles, shelling, and bombs have been indiscriminately launched and dropped on Ukraine. The purpose was not military against the enemy military; the Russian goal was destruction of infrastructure and human population. The free world, led by the United States, initially displayed a timorous attitude for human life by refusing to aid Ukraine in a substantive way for fear of Vladimir Putin escalating the war to a nuclear level. When it became clear that Ukraine was stronger than thought, and Russia was not the menacing threat it claimed to be; the United States poured money into Ukraine and European nations, joined to aid Ukraine in terms of military support and humanitarian assistance. As of July 2022, it appears that Ukraine will endure, and the war could hopefully soon end. That wish began to diminish because the allied response aggravated Putin, and the threat of tactical nuclear warfare by Russia appears to have increased and could become inevitable as Ukrainian forces endured.

As in all cases of governmental decisions and actions, there are unintended consequences. Ukraine is a major provider and exporter of wheat, corn, sunflower oil, and honey, as well as fertilizers, and other vital natural resources. The invasion by Russia, because of its position as a primary source of oil and natural gas, has emerged with self-inflicted injures to its economy; as well as being looked upon with repugnance as a weak country with vile government control and leadership. Russia seems unbothered with a huge supply of nuclear weapons making constant threats to use them. Desperate leaders often do despicable deeds.

The Ukraine War, in combination with the continuing world Covid pandemic, has resulted in worldwide food shortages, national food protectionism, global supply constraints, and worldwide inflation. It will take years for most countries to fully recover from the berserk behavior of Vladimir Putin and the Russian government. Russian warships have blocked access to Odessa, Ukraine, the major Black Sea shipping port where 25 million tons of grain are siloed in the late summer of 2022.

Ukraine produces 25% of the world's wheat and is the source of grain for many European and African countries. Some districts in Africa, especially Somalia, now face starvation as the Russian navy re-enforces its presence in the Black Sea by blocking exporting from Odessa, Ukraine's major seaport. In addition, the war has likely caused a year or more lost in the global response to the impending climate change problems which have been worsened by the war.

Figure 8.6 War in Ukraine

On August 1st, 2022, Turkey brokered a deal with Russia for a ship loaded with 26,000 tons of corn to leave Odessa, Ukraine, and be routed around Turkey to Lebanon. It is unclear if more shipping of grain will be successful, but according to the agreement between Turkey and Russia, 12 grain ships of unknown capacity will be permitted to leave Odessa, for passage to African and European nations.

Partial world grain relief will come from Australia who reported on September 5th that the 2022-year wheat harvest will be a *bumper crop* allowing a major increase in exporting. About the same time, Nord stream the pipeline system delivering Russian natural gas to Germany and other European countries was mysteriously blown-up.

In mid-2023, as Ukraine begins to mount a counter-offensive in the Russian controlled southeastern Russian border region and Crimea. Russia may become more desperate because their economy is suffering from sanctions and lost oil and natural gas income. Desperation could trigger the irrational use of tactical nuclear weapons, or more. If that happens, it could be WWIII during when China, Iran, and North Korea might be involved to aid Russia.

No one knows how long the war could last, some say months, others say years involving other countries. The longer it lasts, the worse it is for climate change and a life of grief for millions of people around the globe. European countries and the United States need to create an exit strategy for Vladimir Putin which will allow him to avoid humiliation or join forces to drive the Russian military into bitter defeat.

It does not make economic or military sense to continue funding an endless war in Europe. If China sees no economic advantage to the war, it may use influence to dissuade Russia. In the United States there is opposition to future financing calling the war a *proxy war* using Ukrainian lives instead of our own to fight our ideological enemy. Mass terrorism against humanity is never condonable or forgivable and be foiled before it starts.

Denial is a save now, pay later scheme.
— Gavin de Becker

CHAPTER 9

MODERATING POLLUTION

Moderating pollution requires a major worldwide collaboration between every country and person on earth. There is no single step, such as battery-powered electric cars that will solve the global air pollution problem. The opinion of many ecologists is that humanity has become addicted to low-cost electrical power consumption, and a tendency to go faster without consideration to the unintended consequences. These mentalities and energy use hypocrisy must change for pollution to become controllable and in balance with natural geologic and atmospheric events.

Humankind has also used earth as a garbage dump, discharging waste into small creeks, to rivers, into lakes, and finally to the oceans. Land has become polluted by mining, factories, logging, industry, sewage, and poor agricultural practices, in addition to widespread carelessness and lack of human discipline. Communication is essential in modern life where technology is ever-changing; and where a need exists for accelerated education of children after two years of global pandemic during which most public schools were partially or fully closed. Major corporations then allowed employees to work at home and only 6% want to return to the offices. Education has always implied a disciplined degree of effort, moderation, and with 7.8 billion people on earth and a growing population, moderation, conservation, and air pollution awareness are essential.

Personal and general conveniences, like cell phone usage, continuous expansion of PC memory and speed, and an absence of education in the physics and economics of power and energy have contributed to the problems. Politics is also a major factor affecting the decisions that leaders of our largest corporations make because profit is necessary to sustain a business. Without profit, businesses fail.

Political whims, favoritism, cupidity, collusion, disconcertion, and a lack of education are the adversaries to air pollution moderation. Each of the major causes of pollution must be addressed by countries in sensible, non-political protocols with carbon-tax penalties assessed to major polluters. The concepts of *externality* and *total power cost* regarding carbon polluters will contribute to climate solutions by discouraging energy waste and minimizing wealth-effect air pollution.

For example, Person A represents low-to-middle class buyers of $25,000 150HP electric cars, and Person B can afford $150,000+, 1000+HP EVS. Person B, because of his/her wealth or attitude, generates air pollution for which the Person A folks are not compensated. They are, in fact, paying for Person B's polluting. The higher the horsepower, the greater the pollution created. That is simply the way that physics works in the world. Therefore, a system of pollution taxes is reasonable with the carbon tax income used to generate alternative clean electrical power.

Similar analogies can apply to energy usages including private planes, large computers, massive ships, railroads, and truck transportation, and fossil-fuel electric power generation. In the year 2020 a study by an analytics company reported that more than one-half of company CEO's used company jet planes for personal and family use.

Elon Musk, a famous engineer who should know better, flew his private jet from San Jose, CA to San Francisco for a meeting less than 30 miles way. He caused air pollution and could have made the roundtrip in less time by driving his Tesla car. Conversely, the developers of carbon reducing technologies should receive pollution tax credits in a bi-partisan system. In government, harsh mandates by a liberal political majority only worsens air pollution problems by dividing the population.

The United States is the world's largest importer, choosing to buy from foreign countries rather than manufacture thousands of goods, even essential metals, whose ores are plentiful in the country. This decision appears more political than practical. Should the United States buy raw and finished metal and chemical products from China, an ideological enemy polluting the globe with coal-fired refining and manufacturing, or make the goods in the United States using alternative energy practices and local workers?

Reducing importing and increasing manufacturing in the United States are issues due for thoughtful bi-partisan reconsideration. One country cannot become green at the expense of other highly polluting countries. That is illogical with a global attitude. Air pollution circulates distributing pollution problems around the world.

Worldwide cooperation is challenging because of cultural, economic, and ideologic differences. In an ideal world, countries with advanced technology and wealth would offer pollution moderating assistance to other countries, especially those with dense populations like China, India, Japan, and others. It may take an unimaginable climate change crisis for that to happen because of economic competition.

This chapter pertains to immediate steps that can be taken to moderate pollution. Moderation entails the conversion of all power generating systems from fossil-fuel to alternative sources of energy. Figure 9.1. is an energy and power source schematic that depicts where and how primary energy conversions require our attention and actions. It is also a graphic summary of the overall air pollution sources and how they prevent natural earth cooling by forming the Greenhouse Effect in earth's atmosphere. The worst polluters are electric power generation using coal and crude oil as fuel.

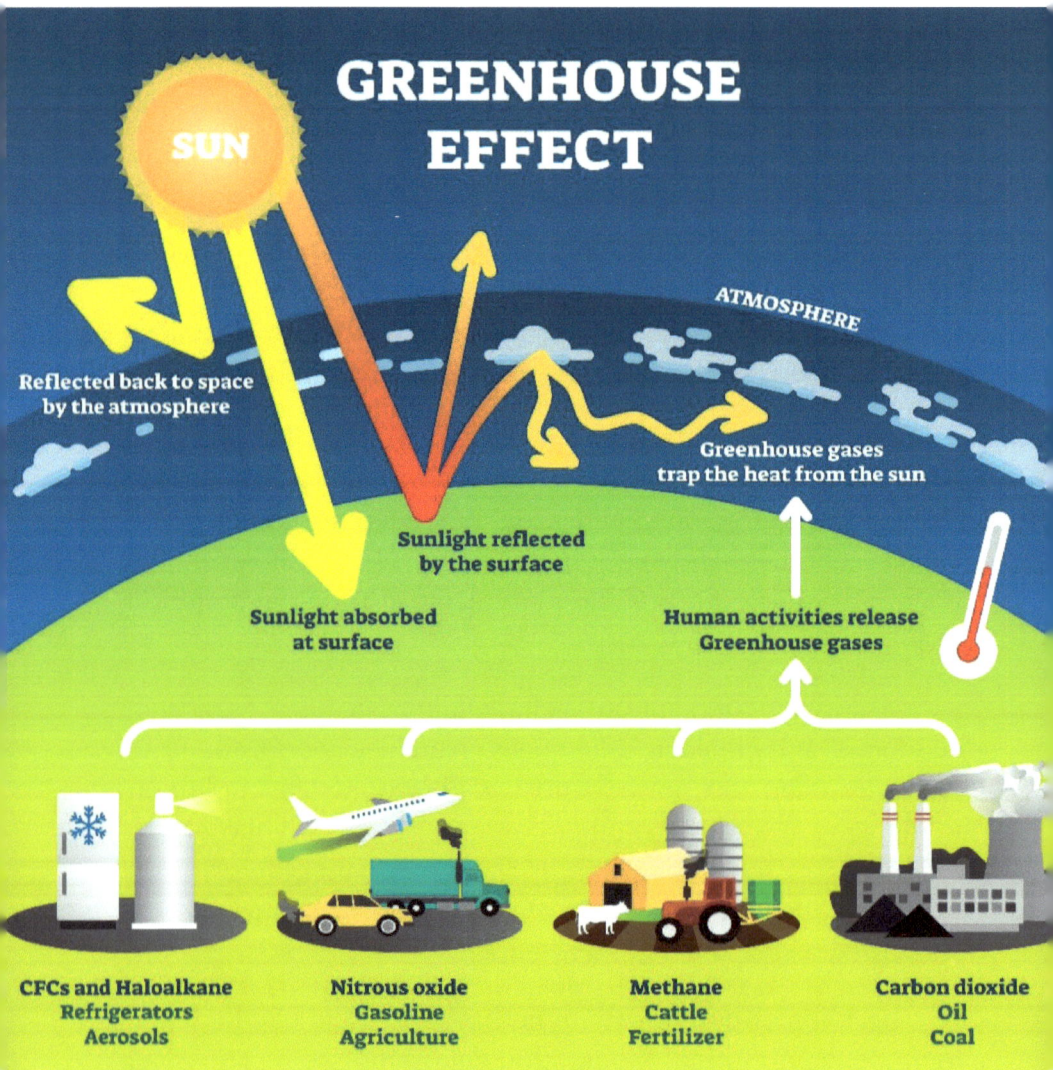

GREENHOUSE EFFECT

SUN

ATMOSPHERE

Reflected back to space
by the atmosphere

Greenhouse gases
trap the heat from the sun

Sunlight reflected
by the surface

Sunlight absorbed
at surface

Human activities release
Greenhouse gases

CFCs and Haloalkane
Refrigerators
Aerosols

Nitrous oxide
Gasoline
Agriculture

Methane
Cattle
Fertilizer

Carbon dioxide
Oil
Coal

Figure 9.1. Pollution Sources and Greenhouse Effect

Cooperation between industries and the government in the United State have a system of carbon debit/credit in place. Carbon is the commodity being exchanged. The details are complicated, suffice it to say that within an industry or company carbon exceeds a limit so a debit is incurred. In other companies or operations carbon is captured and receives a credit. The debit/credits are traded and exchanged like money, tokens, or goods of equal value. This will add a layer of carbon control and fairness in combating climate change.

9.1 Power Generation

Electric power generation is responsible for 32% of the pollution emitted in the United States and tons more in China, India, and Russia where coal stays the most economical and readily available source of energy. The use of coal should be phased out or the power plant be converted to less polluting gas with scrubbers or fuel capture cells deployed on the existing smokestacks. Other alternatives include wind, nuclear, and solar power. Technology companies have designed fuel cell systems that connect to where the CO2 is being emitted. Fuel cell units can capture up to 90% of the stack CO2 and other pollutants and store the carbon in a compressed form in rocks or route it underground to nourish plants and trees.

Coal is found in various forms from high quality anthracite to highly polluting low grade bituminous coal called *lignite*. The highest-grade coal is used to make high-carbon steel with a carbon content up to 2%. High-carbon steel refers to alloys that can be heat-treated for high strength or other qualities such as hardness and ductility. Low quality coal is now used in several countries, especially in China, India, Russia, and other Asian countries as the primary electrical power generator. This needs to change.

India and China have an abundance of labor available to manufacture solar cells and batteries. Power generation using solar cells is a practical electric power generation possibility in parts of these countries. The pollution in Delhi, India is so extreme, that solar power generation for that region would need to be located far away from the central city to receive enough sunlight for solar cell farms that need thousands of hectares of land with sunlight to be economically workable.

Solar cell electrical power generation is low cost requiring moderate capital expenditure and basic design engineering, but not new technology. After a solar cell array is built, it needs only minor maintenance by lower-skilled labor. A means available to every homeowner with adequate sunlight that lowers the power draw on power generation plants is rooftop solar and batteries using inexpensive common salt (NaCl), not high-cost lithium for the cathode. In many parts of the world a home can be powered by the sun with virtually no power drawn from an electric power plant.

There are also several companies that offer components and installation of a complete roof solar system that can be tied to the electric power grid. Tesla Power Wall is one with a cell-phone-type controller enabling the owner to observe and decide when his system has sufficient storage to sell his energy to the power utility. The downside is they are expensive, but in some states zero-interest financing is available with loan payments over a long time. Money saved on the monthly electric bill will pay off the loan. Electric vehicles may help moderate weather change, but the *primary goal* is reducing power drawn from electric power grid, the major cause of global pollution.

A means to ease the drain on the power grid and prevent wide- spread power outages is giant battery packs. FPL uses them in Parrish, Florida, and the Tesla Megapack is in use in Canada and parts of California and Nevada. These are very large 3 gWh batteries that measure 24 ft long x 5 ft wide x 10 ft tall with doors that a person can walk into for servicing. The starting price is over one million dollars! They are worth the price to utility companies to aid in power grid control. Each unit has a bipolar inverter which means that they can either be used to power the grid or be recharged by it or from another power source like solar panels or wind machines.

One gWh is enough energy to power 100,000 average homes for one day based on estimates by the U.S. Energy Information Administration. The multi-purpose units are designed to store power for heavy demand periods from the power grid, emergency standby power, and to aid in managing and regulating stressed power grids. Megapacks are built at the Tesla Nevada Giga Factory near Reno, Nevada.

9.2 Transportation

Movement of people and goods on highways and roads creates the second most pollution on earth today. Transportation by land accounts for about 27% of the world's pollution. When aircraft, trains, and ships are added, transportation is the number one cause of pollution.

Converting to batteries or H2 fuel cells for engines is the most effective long-term remedy if properly conducted. It does not make sense to charge batteries from a fossil-fueled electric power station, nor to rush into the switch from ICE vehicles to electric vehicles. Charging stations are too few, prices too high.

Battery charging must be from renewable energy supplies, not from existing fossil-fuel power plants to reduce pollution. The sources include hydroelectric dams, wind, ocean currents, nuclear, and solar electric power generation. Creating a carbon neutral, clean air environment requires that all major sources of CO2 be eliminated within a reasonable time or be captured and piped to plants and trees growing nearby. Planting trees and plants also helps clean the air.

On September 16, 2022, the U.S. Department of Transportation announced the National Electric Vehicle Infrastructure, NEVI program and released $900 million dollars of the five-billion-dollar fund to states to build non-proprietary EVS charging stations. The majority of these will be built in TX, CA, and FL. The power source is unclear as is the distribution of the profit of income collected.

Private businesses like Tesla and others are doing the same thing as service stations have done before for ICE vehicles. The stations must be powered by the sun, wind, or other renewable sources to be an effective air pollution solution. Private businesses are more efficient than the government doing the same tasks. California has a growing power grid problem and cannot meet the demand now without ordering citizens to not recharge their EVS. When a government owns any power source, it leads to socialism and the ability to control personal choice and vehicle use.

Ocean freighters are significant polluters because they are powered by low-grade bunker-fuel. Economics is the reason for ocean shipping. Importing and exporting basic materials and finished goods is a massive economic engine, but the pollution of ships is difficult to moderate. The best solution in the near-term is for countries to become more self-sufficient producing goods from local natural resources thereby minimizing imports and the air pollution it creates.

Figure 9.2 Bunker-fuel Oil Tanker Under Steam 8.2

Modern countries could insist that visiting ships must be powered by either small nuclear reactors or hydrogen fuel-cell technology. This would be an abrupt political move with short term undesired economic consequences, but it could be phased in on a rising percentage of visiting ship basis.

Converting ships to a cleaner fuel would help, but the cost of shipping would rise. If trade between countries decreased, pollution would moderate. The present huge freighter transport situation requires enhanced carbon capture on land to offset the carbon that ships emit to achieve global carbon neutrality. The situation is unlikely to change within the next few decades.

9.3 Aircraft, Air Freight

Reducing air transportation of imports is another means of air pollution moderation. Heavy planes are powered by kerosene, a highly refined hydrocarbon crude oil product that produces CO_2

clouds that restricts normal surface cooling. There is busy auto and truck traffic to and from airports moving passengers and air freight. As the population increases, air pollution tends to follow global airflight activity.

A company in Germany, Lilium GmbH, is developing a battery- powered jet air-taxi for rapid two-way, city-to-airport service that will help reduce airport emissions. Other companies are experimenting with combining H2 and battery hybrid planes. In the next decade, several non-polluting practical aircraft for both private use and commercial applications will help our air quality.

While alternative energy sources are under development, there are many small steps that can be taken. Methane and other gaseous emissions from oil wells can be captured, and either stored, converted to H2, or be routed underground as plant nutrients.

9.4 Agriculture

All activities on earth contribute to the betterment of humankind if used with consideration to energy efficiency and minimal waste. The same applies to agriculture. Adding 1% seaweed to the feed of dairy ruminants decreases the methane and CO2 in belching by up to 60%. The addition of certain oils and DNA modified bacteria also reduces emissions.

Rotating crops and planting native grasses in pastures is a practice known by farmers for centuries. In many parts of the world, drought is now a problem that is expected by climatologists to worsen. The remedy is to plant crops less dependent on irrigation or plant food in greenhouses. That could pose a financial problem for many farmers because farming tends to be a long-term investment in the land, irrigation, machinery, and technique. A means is needed to ease the burden by zero- interest financing, training, and with an income guarantee. Farmers could face bankruptcy in which the world food supply could suffer even more. Harvesting grains is a polluting operation that emits diesel fuel exhaust and sends dust into the air. The finest particles and CO_2 emitted reaches the atmosphere. Curtailing pollution where it happens, by intelligent control mechanisms, is the best method of moderating pollution. Dust can be captured rather than blown into the air, but farmers cannot afford more costs.

Figure 9.3 Harvesting Grain Creates Airborne Dust

9.5 Manufacturing

Producing a nation's goods for its people or exporting for income is a major energy consuming and power using operation.

Manufacturers now use battery powered tools and machines powered by compressed air. Where it is workable, sun energy, waterpower, or wind can provide the primary energy source for battery recharging. In general, production of all goods and manufacturing can become more efficient with independence from the electric power grid when engineers and management are focused on pollution reduction awareness.

Converting iron ore to steel requires several steps, uses a lot of water, and great amounts of fossil-fuel energy. Steel mills are traditionally built close to coal-fired electric power plants. The combination of a steel mill and coal power plant are the most air polluting processes on earth. Steel making is a highly competitive industry that spends money on research seeking to reduce costs. Boston Metals recently applied a development called Molten Oxide Electrolysis that yields iron from ore emitting vitally no CO2 and releases oxygen locked in the ore. The process appears simple, but the development of techniques and the equipment was complex and expensive.

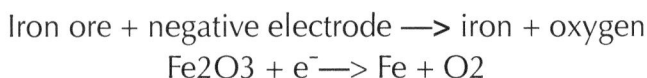

Iron ore + negative electrode —> iron + oxygen
$$Fe_2O_3 + e^- —> Fe + O_2$$

When this technology becomes more universal and other steel mills reinvent how iron ore is converted to iron and steel, a major global generator of air pollution will be eliminated. U.S. Steel pledged to be carbon-free at its Berg, IN plant by capturing 50,000 tons of CO2 per year, the same emitted by 11,000 ICE cars. Others must follow the trend to remain competitive.

9.6 Carbon Capture

Natural and invented techniques can be used to capture carbon both at the source and in the atmosphere. The most common rock on earth is porous mineral rich in iron and magnesium. Basalt is a natural CO2 absorber. Basalt is an igneous (volcanic) rock found in several forms from heavy hard rock to very light glassy pumice that can float on water. The unique characteristic of basalt is its ability to absorb and store CO2 that converts porous basalt into solid rock form over time.

In Iceland, basaltic lava layers are up to several thousand feet thick which makes the land ideal for carbon storage. Experiments have shown that when CO2 is released into the soft basaltic rocks they will harden and the gas in them will remain stored for at least 1,000 years, if not forever. This is a creative example of how dedicated people can reduce air pollution and improve the atmosphere for everyone.

At the opposite tropical parts of earth is the vast Amazon region in Brazil that has served as a massive carbon absorber and oxygen emitter until 2016. Because of deforestation for cattle grazing, lumber products, and agricultural expansion, Brazil became a net carbon exporter. The world lost the greatest source of natural carbon capture, and it worsens every day.

In Africa, jungles have served humanity in a similar fashion, but they are also being cleared for farms as the population grows and food is becoming scarce in Africa and other parts of the world. There is only one way to stop the looming climate change crisis. The atmosphere can no longer tolerate more CO_2. Carbon dioxide must be reduced or captured by industrial means or by planting more trees. The earth is a good place to store CO_2, it holds many mineral oxides that are natural carbon absorbers.

Carbon dioxide can be directly captured from air by absorption, adsorption, membrane gas separation, and other electro-chemical processes. The primary problem with direct carbon capture from surface air is cost and to what altitude it is effective, but the technology methods are advancing. It is far more difficult to capture CO_2 in free air than at the source. Direct Air Capture, DAC uses huge metal structures filled with crystalline sorbent filters that absorb CO_2 when they are cool. After the filters are full, they are heated under pressure and adsorbed out of the filters releasing the CO_2 through underground pipe into rock formations, caverns, or wooded areas for sequestering.

Occidental Petroleum announced in late August 2022 that it was starting construction of a direct air carbon capture project in Texas designed to capture 500,000 metric tons of CO_2 per year by 2023. Details were not released, but the cost will come down as the technology is perfected. Occidental said that it will be carbon neutral by 2050.

Talos Energy, a small independent oil company in Texas, stepped into the carbon capture business by forming partnerships with other oil companies with aggressive goals to clean the air along the gulf coast. Projects will use known and experimental carbon capture techniques.

Habitants on earth must become aware that no part of earth can become a random garbage dump, especially the air that people and animals breathe. Air pollution causes five million human deaths per year. Achieving carbon neutrality and cleansing the air is essential to lowering the number of departed pollution witnesses atop the dark clouds of pollution victims.

Every country has an economic interest in global stability.
— Henry Kissinger

CHAPTER 10

MASSIVE TRANSPORTATION

Massive transportation carriers have been mentioned as major polluters in earlier chapters. The subject is re-emphasized in this short chapter because of the extent of pollution created by moving raw materials and finished goods from producers to users throughout the globe. Ocean shipping is the lowest cost method for moving billions of tons of goods from point A to point B.

The movement of heavy materials, crude oil, fuel, and other goods and products is the cause of 30% of the pollution on earth. The cost of transporting base materials to manufacturers directly affects the price of finished goods. If countries would produce more of their own end products, and use local raw materials instead of importing, air pollution would decline.

The United States is the greatest over-all importer in the world. Trade between China and the United States is the largest between any two countries on earth and it is unequal. Imports from China cost US$2.3 trillion dollars while China imported just US$1.4 trillion in year 2020 before the COVID pandemic. The following three years demonstrated why excessive importing can become disastrous.

The impact of importation resulted in the largest supply chain turmoil ever at the two Los Angeles regional seaports toward the middle of the year 2021 and into late-2022. The effects will be felt for several more years in the form of inflation and parts shortages of semiconductor chips and metals needed to make electric vehicles.

The estimated number of larger ocean ships, mostly gigantic oil tankers and merchant ships, in mid-2020s was 16,000 vessels. They account for roughly 80% of the movement of merchandise from small objects to crude oil, natural gas, ore, automobiles, and heavy machinery. In operation, their engines burn the lowest grade of combustible crude oil. Bunker-oil emits massive amounts of CO_2. At sea, many of these ships dump raw garbage and plastcs into the water.

Figure 10.1 Ships Waiting to Load/Unload

Finished products, cars, food, household items, building materials, chemicals, and base metals are imported from Asia, Canada, and South America rather than be manufactured and mined in the United States. The position of the government administration is that the United States is *greener* by importing rather than producing goods. That position is highly debatable. Pollution occurring in one country has a direct impact on other countries because air pollution is not stationary. It moves constantly in unpredictable cycles throughout the global atmosphere. Pollution in the exporting country circulates and the CO_2 from the transporting ship may be worse than the pollution to manufacture the products a long distance.

Transporting goods and materials on land is the second leading cause of local air pollution. Trucks with 18 large wheels emit CO_2 from diesel engines, as well as unhealthy smog in the form of particulate matter from the tires turning at high speeds causing smog. When worldwide shipping by sea is added with transporting of goods on land, the resulting pollution is not only country issues, but a global polluting condition affecting all humanity and creatures.

Trains are another major source of pollution on land, especially in Canada, China, India, and the United States, as well as some smaller countries. They have supplied economical movement of the heaviest objects, ores, and machinery since the early 1800s. The problem with trains is like ships, they burn polluting diesel oil and are the most economical land mass transporters.

Unfortunately, in the year 2022, the four largest countries are not unified in their approaches to pollution because of political, ideological, cultural, and economic differences. Alliances between friendly countries aid in alleviating pollution in some geographic regions. Worldwide pollution solutions require the cooperation of every country for the common good of all. The likelihood of this happening in the near term is doubtful and there appears to be few joint solutions.

In the meantime, water shipping is still a massive polluter because of the economics of using the lowest grade of bunker fuel to power ships. The remedies, small nuclear reactor propulsion, hydrogen fuel cells, and more efficient engines, are not now economically practical options. The near-term remedy is to reduce shipping by water and on land, and an increased effort to use local natural resources and in-country manufacturing rather than importing goods where labor is available.

The vision for mass transportation is small nuclear reactors for ships and H2 fuel derived from methane for trucking and planes. The major oil and gas companies that were criticized during the COVID pandemic as polluters are science and technology leaders in carbon capture and clean fuel development. They will be the leaders in combating air pollution and in developing alternative fuels.

The most abundant sources of H2 are water and air that are not yet workable H2 sources because of high chemical conversion costs. Methane is the 2nd most air polluting gas in the atmosphere and is the lowest cost source of H2 now. If a future technical breakthrough happens that enables H2 to be economically extracted from water, the world will become a cleaner, healthier planet.

> *A transition to clean energy is about*
> *making an investment in our future.*
> — Gloria Reuben

CHAPTER 11

FOSSIL-FUEL POWERED VEHICLES

A ll fossil-fueled vehicles: cars, trucks, boats, trains, ships, and airplanes emit carbon dioxide and other pollutants. The common term for internal combustion engines is abbreviated ICE. Alternative fuel sources, lithium-ion batteries, other kinds of batteries, H2, fuel cells, and energy forms that mitigate pollution are available now, but are technically difficult and presently expensive. This does not imply that fossil-fuel is worthless. The energy use objective is to reach carbon neutrality through moderation, not by elimination of a vital resource. This chapter discusses the principles of the three most common fossil power devices in vehicles and planes: gasoline motors, diesel engines, and jet turbines, and their advantages and environmental impacts.

Chapter 4 mentioned that the first small gas motors were invented around the late 1880's. A fossil-fuel motor or engine is an internal combustion device that uses one or more cylindrical chambers in which a piston connected to a rotating crankshaft cycling in either a two-stroke or four stroke period. There are two common types of fossil-fuel engines: gasoline and diesel. Both operate by compression of fuel, ignition, and expansion, cycling the piston and its connecting rod up and down. See Figure 11.2 to follow the mechanical steps. The math is straightforward.

The expressions involved are called Boyle's Law and Charles's Law, which relate pressure and temperature to volume in a thermodynamic process.

$$(1)\ \text{Charles's Law: } V1/T1 = V2/T2$$

Steps: With the piston down, V1 is the largest volume and T1 is at the lowest temperature. When a force compresses the volume V2, the temperature rises to T2, and the fuel ignites.

Boyle's Law expands on the expression by considering the pressures in the cycle:

$$(2)\ \text{Boyle's Law: } P1V1 = P2V2$$

Steps: With the piston down, pressure P1 is the lowest and volume V1 is the greatest. When a force decreases the volume, the pressure rises.

Later in scientific development, another physicist combined these quantities in an expression known as Avogadro's Law, and later still the Ideal Gas Law became,

$$(3)\ PV/T = k \text{ where: } k \text{ is a constant.}$$

$$(4)\ \text{Ideal Gas Law: } P1V1/T1 = P2V2/T2$$

where T is in degrees Kelvin that refers to absolute zero, -459.7°F or minus 273°C, a point where enthalpy and entropy reach its minimum value and all motion ends.

To convert Fahrenheit to Kelvin, °F = 1.8 x (°K -273) + 32, and °K = (5/9°F) - 459.7°F or refer to the Figure 10.1 nomogram.

Enthalpy, H is the total heat energy content in a thermodynamic system like an ICE and is equal to the internal fuel heat energy plus PV. In basic math it is written as:

$$(5)\ H = U + PV \text{ where H is the heat energy available to do work.}$$

U is fuel energy. It Is expressed in joules per kilogram of fuel: J/kg. Joule (jowl) is energy or work, J = one watt-sec in engine fuel or 0.2388 calories converted to food energy. The ideal or upper limit of ICE efficiency assumes no entropy change and enthalpy stated as temperature in the Carnot Cycle, Eq (6).

Entropy, S, is Greek for disorder, and is the thermal energy unavailable to perform work. It Is measurable and expressed in joules per unit temperature in Kelvin degrees, J/°K - 1. One Joule (jowl) is energy or work. Kelvin degrees are based on the coldest anything can become, absolute zero.

The simplified equation for the Carnot Effect or Carnot

$$(6)\ n' = W/Qh = (Qh + Qc)\ /Qh = 1 - Tc/Th$$

where Q is heat, W is work, T is temperature, and n' is efficiency expressed as a percentage. Further, h and c are hot and cold temperatures in the reaction.

In words, the Carnot Cycle is the upper limit of efficiency of a thermodynamic engine when converting heat to work as the engine turns. Temperature can be exchanged with pressure and volume as expressed in the Ideal Gas Law in calculations.

Gasoline and diesel fuel are the sources of internal energy. The energy of one ounce of gasoline is 105J = 100kWsec. To convert to horsepower: 1HP = 746 watts, 105/746 = 134HPsec, that represents the instantaneous energy in one second. One gallon of gasoline in a 150 HP car equals about 36 miles driven. ICE vehicles are about 25 to 30% efficient.

(Note: The Ideal Gas Law applies to all fluids from gas molecules to air and heavy liquids in the studies of fluid dynamics and thermodynamics in chemistry, engineering, and the physical sciences. As examples, the lift of an airplane wing, movement of a curveball, and molecular reactions are calculated and expressed using the laws of fluid dynamics and thermodynamics.)

K	°C	F	°C
The boiling point of water 373,15	100	212	100
363,15	90	194	90
353,15	80	176	80
343,15	70	158	70
333,15	60	140	60
323,15	50	122	50
313,15	40	104	40
303,15	30	86	30
293,15	20	68	20
283,15	10	50	10
The freezing point of water 273,15	0	32	0
263,15	-10	14	-10
253,15	-20	-4	-20
243,15	-30	-22	-30
233,15	-40	-40	-40
223,15	-50	-58	-50
213,15	-60	-76	-60
203,15	-70	-94	-70
193,15	-80	-112	-80
183,15	-90	-130	-90
Absolute zero 0	-273	-459	-273

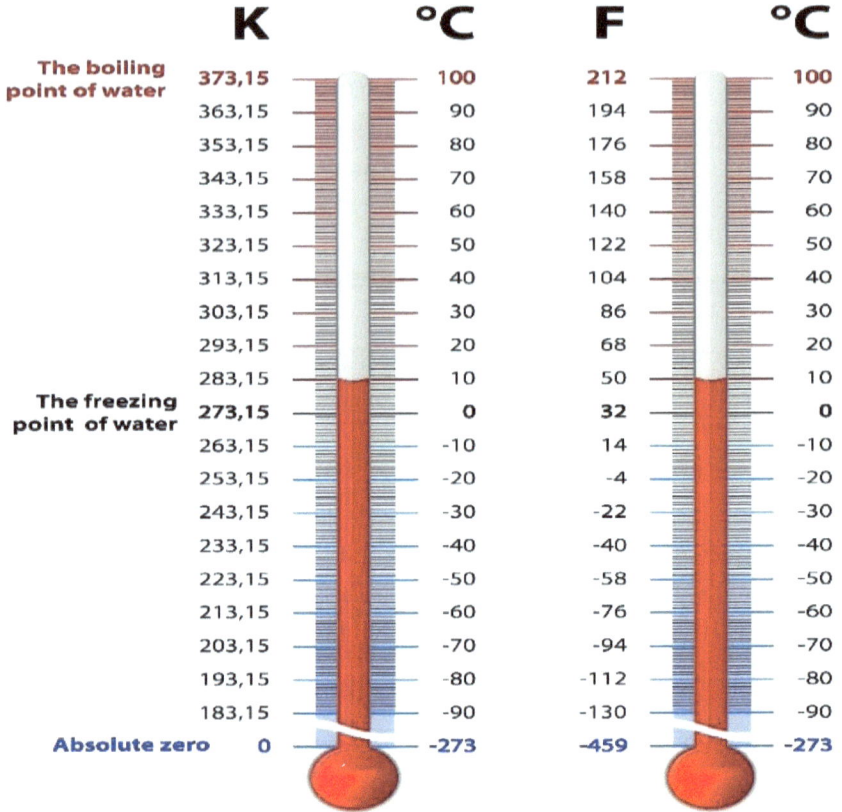

Figure 11.1. Temperature Conversion Nomogram

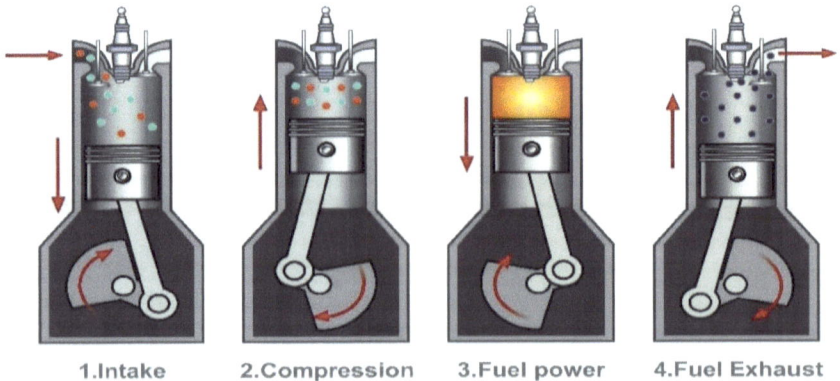

1.Intake   2.Compression   3.Fuel power   4.Fuel Exhaust

Figure 11.2 Internal Combustion Engine Cycle

154

Piston steps in a gasoline engine are: 1) the piston is down and beginning upward by the battery and starter, gas and air are thrust into the cylinder by the intake valve, 2) at the top of piston stroke, a spark ignites the fuel, 3) the piston is forced downward turning the crankshaft, 4) the piston moves upward and the exhaust valve is opened releasing combustion products, 5) a different cylinder brings the piston back down by the crankshaft, and the cycle repeats for each cylinder and piston in unison turning the engine crankshaft that couples to a transmission turning the wheels.

Modern gasoline engines are made with two to twelve cylinders in a wide range of shapes, designs, configurations, and power outputs from about one to several thousand horsepower. Diesel engines are considerably different. They are designed to eliminate the spark plugs and high voltage circuits by dramatically increasing the ratio between the upward and downward compression differences, or ratio. Large power diesel engines and high thermodynamic processes are expressed in mechanics as an inverse percentage, i.e., 12:1.

The high compression ratio P2/P1, in diesel engines, raises the temperature at maximum compression sufficiently to ignite the low octane diesel fuel without a spark plug. The premium grade of gasoline has the highest flash point, or octane rating used in the higher horsepower ICE vehicles. Knocking happens when regular gas is used in a high compression gas engine because the fuel ignites prematurely. Diesel fuel ignites at a much lower temperature. Diesel oil is not refined to the same level as gasoline but is more expensive because the demand for diesel is about 20% that for gasoline cars and trucks.

Engine technology has advanced to enable mass produced gasoline engines in personal use vehicles with 120 HP, or less, that achieve 50 to 65 miles per gallon. Adding catalytic converters to exhaust systems of these engines drastically reduces both carbon monoxide and carbon dioxide. Using these engines in light weight frame and chassis designs make them ideal for low-cost commuting and everyday city driving while the development of low-cost EVS is emerging. In the year 2022, small gasoline cars are considerably less expensive to manufacture and operate than 2023 EVS in 2023 and they are less polluting. The reasons will be discussed in the rest of this chapter and in the following chapters.

The average price of an EVS is about $69,000 and rising as manufacturers struggle with supply constraints, tooling costs, and the learning curve. The same ICE motors, coupled with an alternating current motor, transmission, and small lithium battery, result in a hybrid vehicle that can achieve the equivalent of 90 miles per gallon with very low emissions. As the population continues to grow and cities become more crowded, there is minimal need for high horsepower personal automobiles. In the United States, as weather worsens, the archaic desire for high horsepower and speed will remain until legislation restricts their use; or until a high carbon tax makes them overly expensive and openly frowned upon by the public. Luxury cars will likely be manufactured forever because the demand will continue, and they are the most profitable to make.

Diesel fuel, used mainly for long distance freight hauling, heavy construction equipment like bulldozers, and less so in personal use pickup trucks, is a highly polluting source of energy. The use of low- level oil refinement diesel fuel should be discouraged except in heavy construction, railroad, or mining applications where diesel fuel is essential for high horsepower until an alternative is found. In those applications, there are few alternatives in the year 2022.

Hydrogen, fuel, is a long-term viable possibility. Biofuels made from plants and other organic material have been around for years but were rarely used because crude oil was less expensive. Other strategies are more efficient engine designs, chemical additives to diesel fuel, and the use of catalytic converter technologies that reduce toxic air emissions.

The third type of ICE conveyances are jet engines on planes and less so on some vehicles and other applications. A jet engine can generate tremendous horsepower in the form of thrust in a relatively low weight, small package making them ideal for aircraft. The amount of thrust, F, is determined by the fuel consumed and mass of the airplane, expressed as, $F = ma$.

Lift of the aircraft is a function of velocity and the shape of the wings that create high pressure under the wing and low pressure on top of the wing. Airplanes consume the most fuel and emit the greatest air pollution on take-off and landing to create and maintain lift of the heavy structure. The world's largest commercial airplane is the Airbus A380 that weighs 1.23 million pounds or 617 tons. It flies because of lift and thrust.

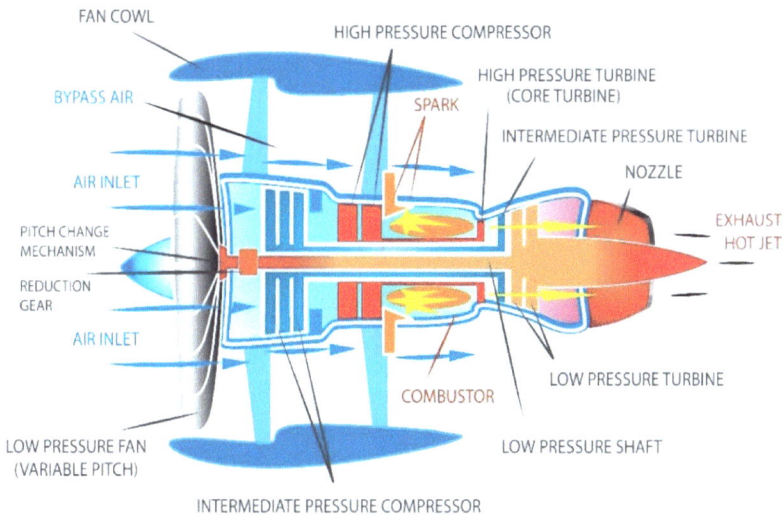

Figure 11.3 Jet Engine Illustration

A jet engine is comprised of several fans connected in series with highly engineered precise turbine blades after which fuel is injected and ignited as shown on the diagram in Figure 11.3. Very high pressure is created and released at the rear of the turbine as thrust. In most cases, kerosene is used as fuel, but a jet engine with modifications can operate on many other fuels including a type of algae and hydrogen. Non-fossil jet fuel is decades away, but hydrogen use is being studied now and its use is perhaps only one decade away from reality.

11.1 Brief Comparison EVS v. ICE

In general terms, there is no global urgency to rush from fossil fuels. The need is to develop and manufacture EVS for use in countries where pollution is the worst because air constantly moves throughout the world. By rushing the transition to EVS, technical errors and manufacturing mistakes will be made. New factory construction, battery technology development, and tooling manufacturing will cause an immediate increase in steel making pollution; the increased demand on fossil-fueled electric power generating plants; and other forms of avoidable air contamination will happen.

The International Energy Association (IEA) has published numerous articles and analysis about comparisons between fossil- fuel and electric vehicles available for readers at, www. iea.org. They say that EVS are about three times more efficient than a conventional fossil-fuel vehicle. On the downside is that the emissions to produce an EVS now is greater than that of a conventional ICE vehicle. An EVS also has a heating and cooling problem. Air conditioning and heating requires about a kilowatt per hour of operation.

Using solar panels on the car upper body surfaces extends range depending on the amount of time the vehicle is exposed to sunlight and the intensity of the light. In the long-term, EVS are a practical pollution solution, but most are years away from perfection, just as the early fossil powered ICE cars were.

In places where EVS are popular and affordable there is an insufficient number of charging stations that are energized by the electric utility power grid. Prior to the September 2022 Labor Day holiday, the State of California issued a warning to all homeowners and individuals to follow these instructions for fear the electric power grid will fail:

1. Do not recharge your EVS from charging stations.
2. Raise your living unit thermostats to at least 78°F.
3. Do not use major appliances, stoves, ranges, etc.
4. Turn off everything that is not vital.

The State of Colorado issued a similar warning and shut down all household thermostats via the electrical power utility's interface network.

Another obvious problem appears to remain in the United States. There are nine EVS personal use models for 2023 that are rated 600 to 1000+HP, and their manufacturers brag that they will go from 0 to 60 mph in less than 2.5 seconds. These were highlighted in *Car and Driver* magazine in the September 2022 issue. These do nothing to mitigate pollution, they are simply sponsoring more of it. The apparent reasons for high HP EVS are human arrogance, promotion by the government, and the cupidity of the car manufacturers. Many people living in the leading economic countries are addicted to high horsepower automobiles and pickup-trucks, often buying vehicles they can barely afford.

What the world needs is inexpensive EVS for everyone, not high-priced toys for the wealthy. The IEA in their articles have suggested that at some point, vehicles should be taxed in accordance with the damage they do to the environment. So far, the author has not seen suggestions elsewhere about taxing EVS based on HP. Taxation is a politically complicated issue. Achieving fairness to everyone is difficult but essential in achieving carbon balance and clean air on earth.

This issue is often decided based for political reasons rather than doing what is the best for most people in a country. Politicians and vehicle manufacturers tend to give people what they want at the highest affordable price whether it is the best for the environment or not. That is why most people are not able to afford an EVS as designed and marketed in the United States as of late-year 2022.

The cost of electric power has risen much higher than the rate of inflation around the globe during and after the COVID pandemic and Russian War on Ukraine. The emergence of lower cost EVS soon is doubtful, and it contradicts the rising price to recharge the batteries. The lower and middle class needing a new car remain in a difficult situation, as the more affluent continue to pollute the air.

> ... *cupidity absorbs all passions and traits.*
> — Samuel T. Hauser

CHAPTER 12

ELECTRIC VEHICLES, EVS

E lectric vehicles, EVS, are not a recent concept. Early versions were designed with a simple motor and built before the 1859 Colonel Drake discovery of oil in Pennsylvania. The earliest models used crude galvanic cells with copper and zinc as the cathode and anode immersed in their liquid sulfates for batteries. The first useable DC motor was made in 1834 by a man named Davenport in Vermont to power his printing press. Battery and DC motor developments evolved, and the first rechargeable lead-acid battery was invented by Gaston Plante in 1859, ironically the same year that crude oil was found in the United States.

The early EVS had a short range, low speed, high cost, and were difficult to recharge because there were no charging stations. They usually had to be returned to the builder when the battery charge ran out. Gasoline engines became popular with modern features and low cost because of mass production and easy access to service stations. As the world began the 20th century, there was nothing in view to slow down production of fossil-fuel cars until World War I.

Factories turned to producing gas and diesel armed vehicles, troop trucks, jeeps, and war planes. Following the First World War, mass production of cars ramped back to personnel automobiles and pick-up trucks. ICE car drivers and taxicabs experienced low- cost mobility and travel around the globe.

Figure 12.1 Simple One Motor EVS Construction
(Occupants sit atop the batteries)

Transportation of people, goods, and materials is a significant cause of global pollution. Replacing ICE engines in cars and small trucks with battery-powered motors is a practical mid- to long- term solution. However, in the near-term, the construction of new factories, refurbishment of old factories, design and fabrication of new machinery, exploration and mining of raw materials, and attendant energy costs **will worsen air pollution**. In the interim, gas-powered vehicles will remain in demand, ideally with lower horsepower, as EVS production plants are built around the world and as companies, or some governments, build more EVS charging stations. In the past, major historical energy use transformations were not aided nor stifled by government actions and political whims, and the people appeared delighted.

In 2002, Martin Eberhard and Marc Tarpenning founded Tesla, the electric car company named after the famous electrical engineer Nikola Tesla. They were among the first men to recognize the pollution problem caused by fossil-fuel ICE vehicles and do something about it. Elon Musk became a co-founder in 2004, and he rose to chairperson and the largest shareholder in 2008. Since then, Tesla has opened manufacturing plants in Newark, California; Shanghai, China; a battery Gigafactory east of Reno, Nevada; two vehicle manufacturing sites in Texas; and an expandable EVS production factory in Germany. Globally, more are planned.

Tesla's first product, called *roadster,* hit the market in 2009. Other models followed: Model S in 2012, Model X in 2018. and the Model 3 in late 2018. The S model was the first electric vehicle to sell one million cars around the world. In addition to manufacturing cars, Tesla invested in solar roof panels, electric grid batteries, charging stations, autonomous car technology, unique computers, and clean energy solutions.

Many difficulties happened along the way. There have been numerous recalls because of engineering errors and manufacturing flaws, and dozens of lawsuits. These problems have not held back Tesla, it continues to grow larger producing more cars every year. On August 15, 2022, Tesla announced completion of its three-millionth EVS – more than 50% of the world total. They were not nearly enough to ease air pollution, but too many for the California power grid.

In addition to becoming the first to mass produce EVS, astute management at Tesla was aware of supply chain problems that would eventually complicate manufacturing of new technology products. They pondered: What will EVS be capable of and look like in 10 or 20 years? What raw materials might become difficult to procure? What human resources will we need? What major components should we purchase from others or design and manufacture ourselves? Tesla asked and answered the right questions, and it became the most valuable corporation in the world in 2021 with new technology and product inventions every year.

Procurement of materials is a major issue with EVS transitions at car companies around the world and in the United States. Vital minerals, Li, Co, Ni, and graphite were not currently mined in the United States and most other countries. Ninety percent of battery anode graphite comes from China and is exported around the world at a price they control. China operates local mines and controls foreign mining and processing of the majority of Li, Co, and Ni mined and processed on earth.

Lithium was not mass produced in the United States either, but at least 10 companies have found and mapped promising Li deposits and applied to the government for slow-to-approve and restrictive mining permission. In year 2021, only a small amount of Li was mined and processed in the United States due to costs, restrictions, and the slope of the learning curve.

Another issue delaying EVS production in the United States is the shortage of semiconductor chips throughout 2021 and through of 2023, caused in part by the COVID pandemic supply constraints, and even more so by the surge of work-at-home personal computer demand. In retrospect, the highly politicized governmental response to the pandemic was grossly mismanaged and miscalculated in many countries, including the United States.

The United States has sufficient Co, Cu, Li, Ni, and other raw materials domestically to build more than two million EVS without importing. Yet apparently for environmental and political reasons, mines are being restricted from producing these metals for policies and assertions by the government that operating the mines and refining processes will worsen the existing air pollution and hasten a climate change crisis. This means that the United States will be competing with other countries for vital materials needed for EVS production. When demand is high for essential materials, the suppliers control the prices resulting in wide-scale inflation. The price of EVS will continue to rise until inflation lowers, or longer.

Figure 12.2 Typical EVS with Body Removed

In August 2022, the International Energy Association reported that 6.6 million EVS and hybrid cars were produced worldwide with one-half manufactured in China. Seventy-five percent of these EVS cars were high-priced, luxury cars meaning that there was limited mass-market participation. In addition, most of the vehicles and their parts were produced in factories that relied on fossil-fueled electrical power generation plants that emit the greatest amount of CO_2 into the atmosphere. The net result is the carbon footprint of EVS vs. ICE cars has gone up!

An EVS uses over twice as much copper as an ICE powered vehicle. By importing the needed metals, mostly from Asia and South America, greenhouse gas emissions from highly polluting ships and diesel trucks abrogates the rationale behind restricting mining and refining in the United States. Major decisions are often made with unintended consequences. Dramatically higher crude oil, diesel, and gasoline prices occurred after the administration decided to shut down oil producers on federal lands in 2021, as a surprise introduction to their intended *climate change agenda* with emphasis on EVS production at any cost.

Electrical vehicles are relatively simple to build compared to fossil-fuel cars with heavy complex ICE motors and dozens of machined moving parts. EVS motors are basically one rotating mass with few parts coupled to a transmission. They are inherently fast accelerating and decelerating, are quiet, do not use energy at rest, and are up to 90% energy efficient.

The quick acceleration feature of an EVS is a pollution enemy in disguise. The manufacturers of high HP EVS publicize the quickness feature to those with a *power-quickness-matters* vehicle mentality. High HP and fast acceleration create pollution because energy is being wasted for no reason. Energy waste is a pollution creator. Haste makes waste is a truism!

Downsides of EVS are cabin heating and cooling during cold and hot weather, and particulate emissions from tires during quick acceleration. EVS batteries need frequent recharging. Fast charging is being promoted now in advertising, lowers the battery recharge cycle life, and replacement batteries are expensive. The cost of operation of EVS is concerning in some countries and regions because the cost of electricity to charge batteries is as variable as the cost of gasoline. The most redeeming feature of EVS is that they emit zero greenhouse gases and will eventually help clean the air. The primary concern is the limited supply of lithium.

The promotion of high horsepower personal EVS is a matter of concern to the IEA. This topic has been touched upon in earlier chapters, but it is worth emphasis. It does not make sense in the world that is frothed with air pollution to waste more energy on expensive toys for the wealthy or for speed show-offs that create air pollution for everyone else. Air pollution occurs in the manufacture of these expensive EVS because of energy wasted to mine more materials, of the carbon released to make higher HP parts, and for mining, producing, charging, and replacing their larger batteries. Speeding around is reckless and air polluting.

These beautiful 1000+ HP cars and personal use pick-up trucks will be designed and built regardless of anyone's opinion. The United States is experiencing the worst highway accident death rate in 13 years, with 38,680 deaths in 2020 despite a reduction in total miles driven because of the COVID pandemic. One-third of these accidents involved drug or alcohol-impaired drivers. Highway speeds have dramatically increased with many gasoline cars and some EVS being capable of speeds over 240 miles per hour. Speed enforcement has lessened in part due to the COVID pandemic, and because of liberal policies toward arresting lawbreakers in parts of the United States. Arrogance and an absence of common-sense accounts for significantly greater air pollution and the loss of human life. Overcoming these attitudes may take years, or until the earth gets even hotter with more deadly hurricanes and tornadoes.

Low polluting electric vehicles will eventually replace ICE autos in the world, and the best way to power them in the short term is with lithium-ion batteries. Material problems based on the year 2022 supply chain disruptions, rampant inflation, and battery-material and semiconductor shortages hindered manufacturing of EVS. Those manufactured and sold in America ended up in TX, FL, and CA where there are electric power grid problems. New technologies are always faced with a steep learning curve.

Lithium-ion batteries in use now in some EVS limit driving distance to under 200 miles. These are fine for limited everyday usage. They can easily be recharged with a simple device using household electric current or from a rooftop solar array setup.

Figure 12.2 Slow Charging an EVS in Home Garage

The most practical method of charging EV batteries is with an expensive roof solar array and garage-installed 110/220 Vac inverter/battery/charger package shown in Figure 12.2., for slow overnight use with a trickle charge. This method ensures a full charge in the morning and extends battery life. An EVS should not be parked or garaged for long periods without a tickle charge because Li batteries self-discharge and lose power left alone.

Tesla Supercharger stations are the industry standard with solar tile roofs and circuitry that enable them to receive power from the electric grid or deliver power to it. There are 20,000 stations globally and 1,728 in the United States. They will recharge the average EVS battery in about 15 minutes at a cost of $20 to $25 per 250 miles or $0.11/mile. Ford and Chevrolet have indicated a desire to have charging stations compatible with all EVS makes and models. All charging stations need to be able to recharge all EVS.

Recharging at home is 33% less expensive, not including the cost of the charging devices. Refueling a mid-price ICE vehicle is about $0.7/mile. The cost of gas and electricity varies widely by country and areas within countries.

Another issue for consideration is battery performance and life in colder climates as compared to warmer climates where both can be severe. Ongoing research and engineering by competing EVS companies to develop the best batteries for various types of vehicles for use in all climates and conditions will result in more efficient and economical power packs. For now, the invention of an ideal, universal million-mile battery exists only as a scientific fantasy. Lucid Air is the first EVS with a 500-mile range.

EVS are equipped with sensors of various kinds that view close stationary and moving objects while they are travelling at highway speeds or during parking. Many allow the driver to press a button for automatic parking, and most EVS will stop the car to avoid a sudden collision. EVS are designed with a *Driver-assist* feature that allows hands-off steering with audible alarms when unusual or abnormal objects are detected. These and other features are either standard or optional on most EVS. The driver must stay attentive while self-drive is engaged, but it relieves driving stress on long highway journeys.

Autonomous driving refers to control of the EVS by itself, unassisted by the driver, or *autopilot* mode. This feature is possible by using lidar technology, an acronym for light detecting and ranging or with video or ultrasonic devices. By using these or similar methods for object detection and avoidance and map algorithms, an EVS can auto-drive to and from locations stored in memory or destinations entered on a keypad. Driver-assist is a safe and convenient attribute, but full autonomy has not been approved except for low-speed city driving. There have been terrible accidents involving Tesla cars in autopilot mode. Many were the fault of the driver, not the car. Further sensor and algorithm development is needed before autonomous driving becomes widespread on major highways in the United States and other counties.

The majority of EVS in the United States have gone to warmer states: CA, FL, and TX. The week of Christmas 2022 presented a more challenging problem to auto-mode when the northern jet stream slammed a historic artic blast of low-pressure air, blinding snow, and -75° F wind chill against a high-pressure zone skidding cars and trucks off highways, massively colliding vehicles unable to gain any traction, and loss of many lives. In winter conditions and on slippery roads, the self-drive algorithms may cause terrible accidents and transportation delays unless they detect surface conditions and disable the feature or are able to compensate for the conditions.

Tesla will offer its full self-drive Tesla Vision option package on September 5, 2022, for $15,000 even as there are lawsuits and government resistance to full autonomy for vehicles. Tesla Vision algorithms will limit highway speeds to 85 miles per hour, still deadly if hitting an immovable object. It is not possible for algorithms to predict every conceivable emergency, event, or situation. Full autonomy continues to improve, but it will never replace or eliminate a good human driver in all conditions on every highway but is better and safer than the average human driver.

On October 26, 2022, the federal government commenced a criminal investigation against Elon Musk charging him personally for statements that he made concerning future capabilities of Tesla Vision auto-pilot mode. This was the same day that Mr. Musk made a visit to the offices of Twitter, the social media company that he bought to stop censoring and promote freedom of speech.

A practical application for full autonomy by other EVS manufacturers is for taxis and personal cars in cities where roads have been accurately mapped and the traffic flow is consistent during the day. Tesla plans fully autonomous, lower-cost robotic cars from its facility in Texas by late 2023.

In time, the EVS will become as marvelous as the first Fords and Chevies that used crude oil for fuel and led the world to weather change. Will it be hydrogen, fuel-cells, or some other new energy technology when the lithium supply on earth is depleted or becomes too expensive?

> *Science can only happen in an*
> *atmosphere of free speech.*
> — Albert Einstein

CHAPTER 13

ELECTRIC VANS AND TRUCKS

The first companies to design and manufacture EVS were Tesla and four Chinese companies. These cars addressed the air pollution problem and weather change, but large cargo trucks were more problematic. Large freight trucks burn diesel fuel and travel five times the distance of the average car. The largest cargo trucks weigh up to 82,000 pounds and need greater power to accelerate and maintain motion. The design of electric trucks is a challenging engineering task in view of weather warming.

Tesla designed prototype cargo tractor called the Tesla-Semi in 2016, that was unveiled in November 2017, in California. The tractor had three carbon-fiber wrapped electric motors with three times the horsepower of a diesel truck. The energy cost claim was 2kWh/mile, far less than diesel semis. The driver sat in the middle of the cab with two large screens replacing rear vision mirrors. It came standard with Tesla Enhanced Autopilot for driver comfort.

Figure 13.1 Tesla-Semi Freight Tractor
Photo Steve Jurvelson, Menlo Park, CA 2017, Wikipedia

Many claims were made about the performance, including a 500-mile range and a cost per mile 20 cents lower than the same diesel tractor and load. The claims gave rise to 745 pre-orders for the semi-truck from well-known freight haulers including Walmart, UPS, FedEx, and PepsiCo, but there were constant news releases about amended specifications and production delays.

One news release concerned a solar recharging package with the claim the truck could be recharged in 30 minutes by its integral solar Mega charger which began tests in Nevada at the Tesla Gigafactory Nevada in late 2020. The battery recharging capacity objective was one megawatt. Whether it was achieved or not is unknown. In 2018, the Tesla-Semi was evaluated with loads on trips between Nevada and California, and longer-range jaunts to J.B. Hunt freight headquarters in Arkansas. Details about the performance and cost per mile were not known when this story was written, but no bad reports were made as of June 2023.

Additional news releases projected cost from $150,000 to $200,000 based primarily on operating ranges from 300 to 500 mile-ranges. The cost difference is clearly based on battery cell materials and their manufacturing costs with hundreds of lithium cells in each battery. Production was to be from the Tesla Giga Reno factory by the end of the year 2022.

There is substantial competition to Tesla from the companies who now build diesel powered trucks in the U.S.A. and other countries. Pollution caused by long distance heavy freight hauling is a major problem that must be addressed to moderate dire weather conditions. The sooner electric freight tractors are available in all countries the better for the world. Air is in constant global rotation.

On October 7, 2022, when work on this book began, Elon Musk announced that the first production run of Tesla-Semi freight tractors will leave the Giga Reno factory in December 2023. They will be delivered to PepsiCo. This was good news and a major step toward addressing the heavy freight CO_2 emission problems. The sooner more are designed and built, the better while hydrogen fuel-cells and new technologies are being developed.

Short distance hauling and city deliveries are an added matter that is now addressed by the major sellers of home-delivery merchandise like Amazon and fleet carriers FedEx and UPS. Vans and small trucks are in or near production by all the major car companies and several newcomers like Nikola, Canoo, Rivian, and the offshore EVS companies. Every step to reduce air pollution is helpful if the method of recharging the vehicles batteries is from renewable clean energy.

Eventually the supply of lithium, nickel, cobalt, and graphite will be more difficult to find, driving their prices upward. The pertinent question is what can replace lithium batteries? The short answer is H2 fuel-cells that may be ideal for large trucks and small ships. Hydrogen fuel-cells are a battery replacement analog with more involved hardware to manage and control the gaseous H2. Their advantages are a huge supply and quick recharging. In the future, H2 fuel cells may become less expensive than lithium batteries. That will be a major achievement to moderate climate change. There is no known element or mixture to replace the lithium in batteries.

In 2021, Clean Technia Incorporated, cleantechnia.com, estimated the cost of lithium batteries at about $175/kWh with 70% or $120/kWh for the cathode whose cost is mostly lithium. The best news is that cost has come down from 2021 as lithium and other expensive battery mineral deposits have been found and mined in more countries.

Tesla managers recently visited Indonesia, a country with 268 million people, 17,000 islands, many gas cars and motor bikes, and rich deposits of nickel, copper, lithium, and tin: minerals that China controls at home or in other countries. The Philippines also has large deposits of gold, copper, and nickel, all vital minerals related to semiconductor chips and EVS manufacturing.

Both Ford and Tesla have negotiated with Canada and Indonesia seeking permission to open EVS manufacturing plants. There will always be a need for the discovery of new mineral deposits and technology as the population expands and as climate problems worsen.

China is the best builder of electric vehicles.

— Elon Musk

CHAPTER 14

ELECTRIC VEHICLE PROPULSION

T he propulsion of land vehicles and sea ships consists of four essential components: 1) An energy source: fuel, battery, or hydrogen fuel cells. 2) A mechanism to convert energy to work, engine or motors. 3) Means to repower the vehicle: refueling system or recharger, and 4) Control devices, both mechanical and electrical. In EVS vehicles, an AC motor replaces the gas engine, and the battery or fuel cells replace the gas tank in an ICE vehicle.

Control and refueling is similar for both cars. The greatest difference is that an EVS emits near zero air pollution, if the recharging method is derived from renewable energy: hydropower, wind, nuclear, or the sun. The difficulties are sourcing the key minerals that are controlled by non-allied countries. The four essential EVS components and procurement of key minerals and materials are covered in this chapter.

14.1 Energy Source

Batteries and fuel-cells were mentioned in earlier chapters with fuel-cells detailed. Lithium battery experiments began in 1912, but progress toward commercial use did not happen until 1980. Then in 1985, the discovery of lithium carbonate as the cathode, and graphite as the anode led to the lithium-ion battery, and chemists began serious efforts to perfect Li-ion cells for a wide range of applications. In 2019, three early Li-ion developmental scientists, John B. Goodenough, M. Stanley Whittingham, and Akira Yoshio received the Noble Prize in Chemistry for their scientific discoveries, pioneering efforts, and achievements.

Lithium is the soft silvery-white mineral that was formed during creation when hydrogen and helium fused as dust. It is the primary cathode element in batteries for EVS cars. The two main types of EVS lithium batteries are:

> LFP Lithium Iron Phosphate, LiFePO4 and
> NCA Nickel Cobalt Aluminum, NiCoAlO2
> Others are in research and development.

The anodes are graphite, and the electrolyte is dense polymer gel. The cell voltage is 2.8 V nominal to 3.6 V at full charge. The cells are connected in series/parallel to achieve the desired voltage and power requirements. Cells in series raise the voltage and connected in parallel current and power are increased.

Total packages store between 120-to-150-watt hours per kilogram, Wh/kg, at a cost of about $120/kWh. The weight of the battery package is more than the weight of an ICE engine and gas tank which makes EVS heavier than an equivalent ICE vehicle and thereby safer. The center of gravity is lower for improved stability and handling, and energy efficiency of an EVS is 70% compared to 25-30% for a gasoline ICE vehicle.

EVS battery package voltage varies with the car design and are between 400- and 600-volts DC. The DC voltage is dependent on the design objectives and characteristics of the vehicle with material cost and long-range emphasis.

14.2 Energy Converter

Figure 14.1 depicts a typical EVS with the body removed exposing the two battery packs, steering devices, motor/transmission mechanics, and blue-colored controller housings. The front of this car is steering mechanisms and driver display of information like most cars. The car rear is where the action takes place. Most EVS are rear-wheel drive with two motors, battery packs, controllers, and transmission gears that convert stored DC electrical energy to rotational force that powers the wheels. Recall that, F = ma, from Chapter 5.2.

Figure 14.1 EVS Underbody

Battery voltage is converted to a three-phase alternating current that generates power using asynchronous or induction motors with a rotating mass (rotor) turned by a stationery magnetic field (stator). The rotor is connected to gears that with electronic controllers make up an electro/mechanical transmission. Speed is a function of the gears, and the voltage and phase control of the AC motors. The ultimate horsepower of the motor unit is measured by the voltage times the current, P = IE delivered to each AC motor expressed in kilowatts minus the losses of the gears times the efficiency of the motor. Horsepower multiplied by 746 converts to kilowatts. For example: using P-power, I-current, E-volts:

$$100 \text{ hp X } 746 = 74.600 \text{ kW}$$
A 100 HP Motor Unit is equal to 74,600 kW.

If the largest stator voltage delivered by the controller is 440 Vac, the current is: 74,600/440 = 170 amperes.

Figure 14.2 shows copper bars (upper center) connected to the fixed stator to manage the high current. If the power delivered to the wheel is 100 HP and the unit is 75% efficient, the motor current would be about 170/0.75 = 227 amperes at full speed. Actual motor units, batteries, and controllers vary by design, size, model, and manufacturer, but all EVS vehicles can be charged with 220 Vac available in homes and from EVS charging stations. Most can be recharged from 110 Vac outlets if they are left on all night and the battery is not overly discharged.

14.3 Recharging EVS

All vehicles require refueling or recharging. EVS batteries are constantly being studied, redesigned, and improved. The objective is to make them recharge more quickly, last longer with more stored energy, and cost less than an equivalent ICE vehicle. The goal is challenging because China is either the primary source or controls much of the supply of the essential battery materials, Li, Co, Cu, Ni, and graphite. This creates a high hurdle to cross as the world transitions to EVS mass production. EVS manufacturers need to produce vital materials at home or from allied countries, not China.

Figure 14.2 EVS Motor, Gears, and Controller

Meanwhile, in most countries, there are not only too few recharging stations, but many of them are also being powered by fossil-fuel electric power generation. An EVS battery recharged with fossil-fuel does nothing to reduce air pollution nor calm weather change effects. Taking one step toward pollution moderation must be made in alliance with or after the primary causes of air pollution, electric power generation and mass shipping, are resolved.

Individuals who can afford the expensive EVS that are produced now will help the cause by installing roof solar panels for home recharging and only frequent commercial charging stations that use renewable clean electrical energy. Unfortunately, the majority cannot afford the 2023 EVS models available in the United States and other countries. They are also over-powered and create air pollution rather than reducing it. The problem relates to supply chain parts shortages of microchips and the dependence on foreign-made metals. Car manufacturers in America cannot make a profit off low-cost EVS models. That means that EVS mass production is several years away and small ICE gas-cars will stay in production.

14.4 Controllers

Modern vehicles all use microprocessor chips and electronically controlled devices. The difference is the primary electrical systems between ICE and EVS vehicles. Lead-acid 12-volt DC batteries are fine for ICE cars, but the electronic system of an EVS is more difficult. Because of the complexity of modern electronic circuits, only the purpose and general arrangement of the electrical system is presented in this book which was written for a wide audience.

EVS uses a high voltage DC battery as the energy storage device rather than the gas or diesel fuel analog in the gas tank. High voltage DC is necessary because AC power cannot be stored in batteries. Higher voltage reduces current draw since the current carrying capacity and cost of copper wire is proportional to wire cross-sectional area ($A = \pi r^2 = 3.142 \times (d/2)^2$. The other reason is that AC asynchronous motors perform most efficiently at higher voltages.

The steps are upon starting, the DC battery is inverted to AC voltage by the controllers. When recharging, the AC source is rectified to DC voltage by the controllers to recharge the DC battery. Recharging can be either 120/240VAC at home or at the charging stations. Slow over-night charging is the best rate for longest battery life without reducing recharge life.

> *There is no genius like the genius of energy and industry.*
>> — D. G. Mitchell

CHAPTER 15

TOTAL ENERGY COST

T otal energy cost is the summation of every energy lost action, or energy made unusable step in the conversion of energy in aa action. It is best explained in a multi-step example: Step 1. A gallon of crude oil is found and extracted from under the ground. Energy is used which makes the potential oil energy worth less than if it were found in a puddle above the ground. Step 2. The crude oil is transported and converted to gasoline in a refinery which uses energy in the process. Step 3. The gasoline is moved using transportation energy from the refinery to a service station. Step 4. Someone buys a quart or so of gasoline refined from a gallon of crude oil to fuel a motorcycle. Step 5. The motorcycle transports the owner, but CO_2 is emitted. What is the total energy cost? The short answer is how far the motorcycle goes minus the amount of pollution emitted.

Total energy cost equals the energy used in Steps 1 through 5. In addition, human energy was used in each step. Human energy originates from the food calories consumed and is lost in each step taken. It also takes energy to grow the plant and animal food and get it to the market. During each of the energy steps, CO_2 and other pollutants are emitted into the surface air and atmosphere. The steps are *carbon-footprints*. Multiply every action for all humans' times everything with mass and the result is total energy cost.

If a battery had been used to power the motorcycle or a vehicle of any kind, it would have taken multiple energy steps to mine and acquire the various materials, produce the battery, test it, and install it on the motorcycle or in the vehicle. It takes energy to use all the energy forms on earth regardless of the chemistry of the material or product. Lithium and other chemicals for EVS batteries can be considered as analogs of fossil-fuels.

Deciding the best use of energy is a serious and complex assessment. *Decarbonization* requires that energy production be low-polluting, renewable, and sustainable if the world is to become carbon-neutral or balanced. While it may be more efficient to develop horsepower from a battery and electric motor than from a fossil-fuel vehicle, a high HP vehicle of any design will always use more energy than the equivalent lower HP vehicle. Energy output equals energy input times efficiency and is always less than 100%.

The total energy cost in the foregoing is elementary. A similar analysis of major energy decisions should be performed regarding all alternative energy transformations.

Typical questions and issues industrial leaders and their scientists and engineers, and the governments, should be asking:

1. What are the primary raw energy forms, raw materials, factory, and human assets? Is there sufficient electrical power for the future? How will it be generated?

2. What are the problems in obtaining essential minerals for goals and programs that we do not have in our country?

3. How does existing infrastructure relate to reducing pollution? Can coal power generating plants be modified?

4. What are the total EVS project energy costs: mining, refining, processing, transporting, training, facility refurbishment, new construction, and tooling costs?

Is there adequate housing, water, and food energy for employees?

5. Will we continue using polluting electrical power (coal) to produce our low-emission products? Are there alternatives for each step in our energy converting processes?

6. How do we address our major polluters, power generation and transportation? How do we discourage air pollution as a choice? How do we educate in an atmosphere of ideologic and political division?

7. Why should not the government tax excessive horsepower for personal vehicles? Excessive horsepower pollutes! This is called *common sense*, often ignored by the arrogant elite. Decisions must be based on what is best for everyone, not specific groups as the earth continues to overheat with more people using energy that emits air pollution.

8. How and to what extent will the government finance alternative energy and pollution abating projects? What will the government control, regulate, and restrict? Politics is a major uncertainty on earth.

9. Are we taking the correct steps to prevent the next world war or bring it to a quick end if it happens? Do we have enough oil stored in the SPR for defense and emergency use with the possibility WWIII?

10. When will Americans become used to smaller, lower HP vehicles, if ever? How do we educate the population that bigger and faster means pollution and waste? Human adaptability may become a necessity rather than a choice.

11. How will we adapt to decreasing water and food shortages, more inflation, a recession, fuel insecurity, lost years of child education, the next pandemic, weather extremes, and possible WWIII?

Readers can add their own questions and thoughts to the list. The goal is to achieve balance, also known as *carbon neutrality*, and peace on earth. The sun and earth do their parts, humankind must do theirs without hindering natural processes. Consideration of others and common sense by reducing speed, minimizing power waste, and being more efficient and healthier human beings will lead the way to a cleaner, more peaceful world.

Energy is used every time someone breathes, even in sleep, or when anything moves. Without proper nutrition every living organism is at risk of dying. The same principle applies to earth. It is the duty of everyone to be energy aware and pollution conscious to nourish, support, and protect where we live, breathe, eat, and acquire and utilize natural resources for our human energy.

> *Learning is not compulsory…Neither is survival.*
> — W. Edwards Demin

CHAPTER 16

CONSERVATION OF ENERGY

T he Law of Conservation of Energy states that energy can neither be created nor destroyed, only transformed, or transferred in a closed system. Pollution is the result of overuse and misuse of the energy resources on earth and the inefficiency in converting energy to work. The solution is to use alternative power generation, research alternatives, achieve greater conversion efficiency, and for everyone to learn how to minimize power usage without lowering quality of life. In 2020, the average person in the United States used 29 kWh of power per day, the highest on earth for a large country. In India the usage was 6 to 7 kWh per day per person. Most people in India do well without air conditioning or high horsepower vehicles they do not need or cannot afford. Both countries have work to do.

Planet earth has reached a critical point in time where power usage has accelerated beyond the known power technology and resources to generate more energy as the population continues to grow. In addition, common sense and leadership appears to be at a record low. The United States has a limited supply of crude oil estimated to last for about seven years without moderation, new oil field discoveries, or oil importation from other countries, yet we continue to manufacture 500+HP cars for commuting and trips to the grocery store, doctor visits, and gas station fill-ups.

Many Americans have an automobile for every member of the family old enough to get a driver's license. There has been no leadership action taken to convince households to switch from incandescent to light emitting diode lighting. In fact, the author cannot recall any president since World War II ever addressing the country urging energy use moderation. It is simply not a politically viable message that implies something is not going well in the administration. Politicians are experts at telling people what they want to hear. So, what can you do to address impending climate change and energy problems?

The answers are simple. Just be energy aware and use common sense and encourage and promote energy awareness with others. This list is not intended to cause climate alarm nor imply the need for a lowered quality of life. They are energy use suggestions that may reduce your energy costs and improve your quality of life.

1. Buy LED light bulbs. LED lighting is at least five times more energy efficient than incandescent bulbs. LEDs are more expensive, but they last 12 to 15 times longer with the prices coming down as mass production Increases. Turn off unnecessary lights, fans, and appliances.

2. Use less water, avoid long showers. Many people do not need a shower every day. Turning down the water heater temperature extends the life of water heaters. Plant less thirsty lawns and landscaping. Use native plants and rocks to beautify your home and property. Fresh water is becoming scarce, perhaps scarcer.

3. Learn how to acclimate your body to be comfortable in both warmer and cooler temperatures. Setting household thermostats just two or three degrees lower in winter and higher in summer will reduce the amount of power used and lower your energy bill.

4. Plan short trips and purchases to avoid wasting gas and power. Don't hoard, buy what you need, do not waste food. The food supply will change as weather worsens and as farmers face high diesel prices and labor shortages.

5. Do not buy horsepower that you will never need. Ninety-five percent of trips are made close to home according to the International Energy Association. There is nothing wrong with driving a modest car to the market.

6. Slow down, you will get there just as soon, and more safely by not speeding to an intersection and stop lights, and then applying hard braking. Highway death is at a record high.

7. Recycle paper, cardboard, most plastics, and metals, and donate acceptable reusable items to charity. Use waste disposal service and do not litter.

8. Be involved in local politics if you are able. Vote for candidates who have logical energy platforms and who understand the need for alterative power generation regardless of party affiliation. Liberal ideology benefits few, excuses crime, increases taxes, and causes inflation.

9. Educate your children and those of others if you have an opportunity. Education is the key to solving earth problems. It starts before the first grade with *haste makes waste*.

10. Help your school children restore their knowledge level to grade ratio lost during COVID-19 lockdowns. Teach the values that you wish to instill in them. Learn what they are taught, and support traditional school board members.

11. If you must travel, consider the least polluting means.

During the last three years, the world has endured a pandemic, supply constraints, food insecurity, political division, a costly war, inflation, broken borders, and Russian-China aggression. Much of the world and the United States may be feeling the greatest degree of anger, anxiety, crime, drug use, fear, grief, and stress over abortion, border, climate, food, fuel, gangs, gender, inflation, policing, racism, sexuality, and socialism in history. If these are causing you stress, here are some thoughts about life as expressed by notables of the past.

Present fears are less than horrible imaginings.
— William Shakespeare

The universe doesn't allow perfection.
— Stephen Hawking

The only cure for grief is action.
— G.H. Lewis

Fear not for I am always with you.
— Isaiah 41:10

CHAPTER 17

CONCLUSION, AND A
NEW BEGINNING

Human population growth necessitated humankind to develop novel methods of harvesting food, and in developing advances in manufacturing techniques. These needs led to a prompt increase in the use of the earth's energy forms. The period from 1770 until the present day is known as the Advent of Mass Production. It can also be called the Dawn of Pollution, during when the egocentric control of the most readily available energy forms: *fossil-fuel*, and human cupidity, overcame discipline and sensibility related to resource management on earth.

Waste in the form of smog, particulate air, and extensive damage to earth's atmosphere and environment was the result. Garbage has been disposed of everywhere from the tiniest creeks and hiding places to the most extreme parts of earth, Mt. Everest, the highest, to the Mariana Trench, the lowest. The greatest part of earth, the oceans, and its food resources, have been in steady decline as waters were strewn with garbage, and marine creatures became poisoned and choked with non-biodegradable plastics.

In 2018, Greta Thunberg, a 17-year-old Swedish environmental activist, sounded the weather warming alarm, and it triggered global panic to save the earth from the *climate change crisis,* a political misnomer designed to frighten the public. She was nominated for a Noble Prize for three consecutive years. The climate has not changed because of natural geologic and atmospheric alterations; pollution has caused extreme weather warming and cooling, and other health problems around the globe. What should we do?

The answers are known, do not politicalize or over-regulate major earth problems. Listen to science and futurist industrial leaders who understand the problems and have been creating solutions for centuries. It is essential to decrease the use of fossil fuel and replace it with renewable energy for electric power plants. Fuel cells can convert methane to H2, and CO2 can be captured from gas-fired power plants for land and water transportation. Conserve vital natural resources, find alternatives, continue to research materials and processes, and combat pollution through education.

Wars, pandemics, famines, and catastrophes have inspired creative humankind to research, invent, engineer, and produce solutions for the natural and human caused disasters of the world. We studied the smallest organisms on earth, bacteria and viruses that kill millions of humans, animals, and plants. Solutions take time, but humanity is ineptly behind in natural resource management, pollution control, and air quality on earth.

When a confluence of problems arise, time can appear belated. However, scientific research is a continuous process of examining the smallest and largest creatures on earth for clues to solving human problems from medical research to climate solutions.

Researchers are now studying why a giant tortoise can survive for 140 years eating only once a year, while a giant manta ray will die if not eating continuously. At the opposite end of the scale, butterfly wing studies have steered us to solar nanofibers that are able to auto-focus on the sun. Other butterfly wing research has led to uniquely hydrophobic nanopillar materials relevant to climate solutions, and nanostructure material for human skin, bones, and eye surgery. The life of a butterfly is one to two weeks!

Winning the air pollution battle and managing extreme weather is on the side of humankind if we can get going before more dark clouds of pollution victim ghosts appear in the sky.

Getting a good start, trump done perfectly.

Conversions, Equalities, and Energy Events

Conversions

1 watt = 1-volt times 1-ampere
1kwh = 3.6×10^6 J
1 joule = 1watt-second
1 ounce = 28.4 grams
1 ft.lb = 0.737 J
1 calorie = 4.18 J
1 gallon = 128 ounces or 4 quarts
1 gallon = 4.2 liters
1 kilogram = 2.2 pounds
1 inch = 2.52 cm
1 meter = 39.54 inches
1 St Mile = 5,280 feet
1 sq. mile = 640 acres
1 acre = 43,560 sq. ft.
1 hectare = 2.47 acres
1 nautical mile = 6,076 feet
1 mile = 1.61 kilometers
1 cc = 1 milliliter (H2O)

Equalities

$1eV = 1.6 \times 10^{-19}$ J
1 lb. gasoline $= 2.2 \times 10^7$ J
1 lb. U235 $= 3.7 \times 10^{13}$ J
1 metric ton $= 1,000$ kg. or 2205 pounds
1 nautical mile $= 1.15$ statute miles
1 light-year $= 5.9$ trillion miles
1 barrel of oil $= 840$-mile car trip
1 HP $= 746$ watts
$\quad = 550$ ft-lb/sec or 33,000 ft-lb/min
Speed of light $= 186,300$ miles per second
Circumference of earth $= 21,600$ Nt Miles
$\quad\quad\quad\quad = 24,840$ St Miles

Energy Events[1]

Creating our Universe 10^{68} J
Mass-energy of Sun 2.2×10^{47} J
Supernova 10^{44} J
Earth Orbit 10^{39} J
Daily Energy of Sun 3.3×10^{21} J
Russian Nuclear Bomb Test 2.1×10^{19} J
1,000 mW Power Station 10^{16} J
Cat 2 Hurricane/second 6×10^{14} J
Thunderstorm 10^{15} J
Transatlantic Jet Flight 10^{12} J
One Ton of Oil 42×10^9 J

---

1   Source: *Energy, Magnitudes of*, Wikipedia

PERIODIC TABLE OF THE ELEMENTS

PERIODIC TABLE OF THE ELEMENTS

Legend:
- Alkali metals
- Alkaline earth metals
- Transition metals
- Post-transition metals
- Metalloids
- Nonmetals
- Halogens
- Noble gases
- Lanthanides
- Actinides

IA	IIA	IIIB	IVB	VB	VIB	VIIB	VIIIB			IB	IIB	IIIA	IVA	VA	VIA	VIIA	VIIIA
1 H 1.0079																	2 He 4.0026
3 Li 6.941	4 Be 9.0122											5 B 10.811	6 C 12.011	7 N 14.007	8 O 15.999	9 F 18.998	10 Ne 20.180
11 Na 22.990	12 Mg 24.305											13 Al 26.982	14 Si 28.086	15 P 30.974	16 S 32.065	17 Cl 35.453	18 Ar 39.948
19 K 39.098	20 Ca 40.078	21 Sc 44.956	22 Ti 47.867	23 V 50.942	24 Cr 51.996	25 Mn 54.938	26 Fe 55.845	27 Co 58.933	28 Ni 58.693	29 Cu 63.546	30 Zn 65.39	31 Ga 69.723	32 Ge 72.64	33 As 74.922	34 Se 78.96	35 Br 79.904	36 Kr 83.80
37 Rb 85.468	38 Sr 87.62	39 Y 88.906	40 Zr 91.224	41 Nb 92.906	42 Mo 95.94	43 Tc (98)	44 Ru 101.07	45 Rh 102.91	46 Pd 106.42	47 Ag 107.87	48 Cd 112.41	49 In 114.82	50 Sn 118.71	51 Sb 121.76	52 Te 127.60	53 I 126.90	54 Xe 131.29
55 Cs 132.91	56 Ba 137.33	57-71 La-Lu	72 Hf 178.49	73 Ta 180.95	74 W 183.84	75 Re 186.21	76 Os 190.23	77 Ir 192.22	78 Pt 195.08	79 Au 196.97	80 Hg 200.59	81 Tl 204.38	82 Pb 207.2	83 Bi 208.98	84 Po (209)	85 At (210)	86 Rn (222)
87 Fr (223)	88 Ra (226)	89-103 Ac-Lr	104 Rf (261)	105 Db (262)	106 Sg (266)	107 Bh (264)	108 Hs (277)	109 Mt (268)	110 Uun (281)	111 Uuu (272)	112 Uub (285)	114 Uuq (289)					

57 La 138.91	58 Ce 140.12	59 Pr 140.91	60 Nd 144.24	61 Pm (145)	62 Sm 150.36	63 Eu 151.96	64 Gd 157.25	65 Tb 158.93	66 Dy 162.50	67 Ho 164.93	68 Er 167.26	69 Tm 168.93	70 Yb 173.04	71 Lu 174.97
89 Ac (227)	90 Th 232.04	91 Pa 231.04	92 U 238.03	93 Np (237)	94 Pu (244)	95 Am (243)	96 Cm (247)	97 Bk (247)	98 Cf (251)	99 Es (252)	100 Fm (257)	101 Md (258)	102 No (259)	103 Lr (262)

About the Author

Patrick was born in Molalla, OR, a town of 100 people south of Portland. He attended parochial schools from grades 6 to 12. In 1951, he joined the USCG and agreed to take duty in Korea. After that he was in college in OR and CA studying U.S. history, philosophy, and science. Then he signed a contract with G.E. Atomic Power Division in 1959 as a technical writer. His proposal for the Hanford Nuclear N Reactor neutron flux monitoring and safety-system was issued and G.E. asked him to design the 29-channel system. The design and he were toasted by the Instrument Society of American in 1963 at the Hanford Atomic Energy Works in Washington state.

In 1963, he was assistant professor of engineering at San Jose State University and director of product development at Magnuson Engineers and awarded seven patents in food processes. Bob Magnuson died, and he returned to G.E. as manager of nuclear instrument marketing in Europe. Nuclear fell out of favor and he started SimCo in Medford, OR, to design log measuring technology. A CA defense contractor bought SimCo in 1975.

Patrick spent seven years designing and building custom homes in OR and NV with his son prior to his microprocessor interests. In 1985, he began Navigation Technology International to fill the need of small boat and airplane users for low-cost, full function NAV computers. His PL-99 was the world's first hand-held NAV computer, followed by a GPS computer he licensed to eight clients in Australia, Brazil, Canada, China, Singapore, and the U.S.A. for precise global satellite navigation.

Simmons is a licensed complex aircraft pilot and owned a Cessna Cardinal-RG airplane and two seagoing sailboats. He lives in Bradenton, Fl, where he huddled while Hurricane Ian, FL's worst ever, scuttled.

GPS NAV Computers and Circuit Boards Testing
Photo: Editor, Portable Design Magazine Carson City, NV January 21, 1996

PL-99, World's 1st Handheld Loran-C NAV Computer C
Photo: Bill@loran-history.info
(Still sold on eBay as a historic souvenir)

Old designs and technology are always replaced with the latest.

www.ingramcontent.com/pod-product-compliance
Lightning Source LLC
Chambersburg PA
CBHW040936030426
42335CB00001B/3